Also from Westphalia Press
westphaliapress.org

War Between Japan and Russia

War Between Japan and Russia: With Historical and Descriptive Sketches of Russia, Siberia, Japan, Korea and Manchuria

Westphalia Press
An imprint of Policy Studies Organization
1527 New Hampshire Ave., NW
Washington, D.C. 20036
info@ipsonet.org

ISBN-13: 978-1-63723-200-2

Cover design by Jeffrey Barnes:
jbarnesbook.design

Daniel Gutierrez-Sandoval, Executive Director
PSO and Westphalia Press

Updated material and comments on this edition
can be found at the Westphalia Press website:
www.westphaliapress.org

War Between Japan and Russia

With Historical and Descriptive Sketches of Russia, Siberia, Japan, Korea and Manchuria

---

by Richard Linthicum

Contributions by
Trumbull White

WESTPHALIA PRESS
An imprint of
Policy Studies Organization

SILENCED.
An Episode in the Defense of Port Arthur.

GENERAL NOGI.

The Man to Lead the Japanese Army to Victory at Port Arthur.

GENERAL STOESSEL.

The Russian Officer who Commanded the Garrison at Port Arthur.

WAR BETWEEN
JAPAN AND RUSSIA

THE COMPLETE STORY OF THE DESPERATE

Struggle Between Two Great Nations

WITH

Dominion Over the Orient as the Tremendous Prize

Describing and Picturing the Mighty Conflict, the Hitherto Resistless March of Russian Force Across Asia, and the Advance of Japan into a Position of World-Power Among the Nations

WITH HISTORICAL AND DESCRIPTIVE SKETCHES OF

RUSSIA, SIBERIA, JAPAN, KOREA AND MANCHURIA

CONTAINING MANY MAPS AND ILLUSTRATIONS

BY

RICHARD LINTHICUM

Of the Editorial Staff of the Chicago Chronicle, Author of "Boer and Britisher in South Africa," etc., etc.

WITH COPIOUS INTRODUCTION AND SPECIAL CHAPTERS

BY

TRUMBULL WHITE

War Correspondent and Historian, Author of "War in the East" (a history of the Chinese-Japanese war of 1895), "Our War With Spain," etc., etc.

Dedicated

to

The Heroes of War

and

The Lovers of Peace

Chronology of Russian Aggression in Asia Which Resulted in the Russo-Japanese War

1860—Russia annexes extreme eastern part of Manchuria, thus securing the port of Vladivostok.

1881—Russia occupies more Manchurian territory.

1885—Russian colonies established in Manchuria.

1891—Czar Alexander III. issues a rescript for the building of the trans-Siberian railroad.

1895—Treaty of Shimonoseki, making peace between China and Japan. France, Germany and Russia compel Japan to withdraw her claims as to portions of the Liaotung peninsula, and Japan yields to superior force.

1896—Private treaty between China and Russia, the latter promising to defend China against the world; Manchurian railway agreement perfected.

1897—Germany gets possession of Kiau-Chou. Russia takes the opportunity to seize Port Arthur and Talien-wan.

1900—Boxer rebellion gives further opportunity to Russia to fasten her grip on Manchuria.

1901—Protests by Japan and other powers as to Russian encroachments on Chinese territory.

1902—Treaty of alliance between Great Britain and Japan is signed Jan. 30. China-Russian treaty signed April 8 for the evacuation of Manchuria by Russia, and declaring the province to be an integral part of China; China's commercial treaty with Great Britain signed Sept. 5.

1903—Russia fails to keep pledge to evacuate Manchuria, and April 23 demands new conditions as to the carrying out of the Manchurian treaty.

June—Japan and the United States object to the new conditions, and Russia later says she has demanded no new conditions. Russia begins movements on the Yalu River under pretense of taking up timber concessions.

July—Japan protests and presses Korea to open the port of Wiju to foreign trade. Great Britain and the United States concurring in the demand. Pavloff, Russian minister to Korea, makes threats.

August—Japan proposes a modus vivendi the 12th, and negotiations are opened. Japan presses Russia for fulfillment of all treaty conditions.

September—Russia gives new pledge that New Chwang and Mukden would be evacuated Oct. 8, but later demands grants for landing places and military post stations, and Japan evinces resentment.

Oct. 8—China signs commercial treaties with Japan and the United States.

Oct. 15—Japan agrees to recognize dominant interests of Russia in Manchuria, but wants Chinese sovereignty unimpaired. It also demands recognition of Japanese influence in Korea and the opening of Yongampho and other ports, besides a neutral zone on both sides of the Yalu River.

Oct. 30—Reoccupation of Mukden by Russia is reported.

December—Japanese diet opens the 10th and the government is censured for the inadequacy of its measures.

Dec. 17—Russia makes reply to Japan's last note, but it excludes Japan altogether from the Manchurian question.

Dec. 21—Russia is asked to reconsider its reply.

Dec. 28—Emergency ordinance is issued in Japan, giving the government unlimited credit.

Dec. 29—High court of war is held, with the emperor in the chair. Admiral Togo is put in command of the fleet; he captured the Taku forts in 1900.

1904—Jan. 5—Japan sends final note to Russia, intimating that her demands must be acceded to or war will result.

Jan. 8—Review of troops is held at Tokio and great enthusiasm is shown.

Jan. 15—Russia delays answer and Japan shows impatience.

Jan. 30—Minister Kurino makes inquiries at St. Petersburg as to when a reply to Japan's last note may be expected.

Feb. 3—Russia's reply is sent to Tokio, but proves unsatisfactory.

Feb. 5—Baron de Rosen, Russian minister, is notified at Tokio that Japan intends to sever diplomatic relations with Russia.

Feb. 6—Minister Kurino, Japanese envoy, demands his passports at St. Petersburg and receives them. Baron de Rosen is notified to quit Tokio.

Feb. 8—First shot of the war is fired near Chemulpo, Korea, in the afternoon, and three Russian warships are blown up by Japanese torpedoes in Port Arthur at midnight.

Feb. 9—War declared.

Japanese Imperial Rescript Declaring War on Russia

We, by the grace of heaven, emperor of Japan, seated on the throne occupied by the same dynasty since time immemorial, hereby make proclamation to all our loyal and brave subjects as follows:

We hereby declare war against Russia, and we command our army and navy to carry on hostilities against her in obedience to their duty with all their strength, and we also command all our competent authorities to make every effort in pursuance of their duties to attain the national aim with all the means within the limits of the law of nations.

We have always deemed it essential in international relations, and have made it our constant aim, to promote the pacific progress of our empire in civilization, to strengthen our friendly ties with other states, and to establish a state of things which would maintain enduring peace in the far east, and to assure the future security of our dominion without injury to the rights or interests of other powers. Our competent authorities have also performed their duties in obedience to our will, so that our relations with all the powers had been steadily growing in cordiality.

It was thus entirely against our expectation that we have unhappily come to open hostilities against Russia. The integrity of Korea is a matter of the gravest concern to this empire, not only because of our traditional relations with that country, but because the separate existence of Korea is essential to the safety of our realm.

Nevertheless, Russia, in disregard of her solemn treaty pledges to China and her repeated assurances to other powers, is still in occupation of Manchuria, has consolidated and strengthened her hold on those provinces, and is bent upon their final annexation. And since the absorption of Manchuria by Russia would render it impossible to maintain the integrity of China, and would, in addition, compel the abandonment of all hope of peace in the far east, we were determined in those circumstances to settle the question by negotiations and to secure thereby permanent peace.

With that object in view our competent authorities by our order made proposals to Russia, and frequent conferences were held during the last six months. Russia, however, never met such proposals in a spirit of conciliation, but by wanton delays put off a settlement of the serious questions, and by ostensibly advocating peace on one hand, while on the other extending her naval and military preparations, sought to accomplish her own selfish designs.

We cannot in the least admit that Russia had from the first any serious or genuine desire for peace. She rejected the proposal of our government. The safety of Korea was in danger, and the interests of our empire were menaced. The guarantees for the future which we failed to secure by peaceful negotiations can now only be obtained by an appeal to arms.

It is our earnest wish that by the loyalty and valor of our faithful subjects, peace may soon be permanently restored, and the glory of our empire preserved.

MUTSUHITO.

Russian Supreme Manifest Declaring War on Japan

By the Grace of God we, Nicholas II., emperor and autocrat of all the Russias, etc., make known to all our loyal subjects:

In our solicitude for the maintenance of peace, which is dear to our heart, we made every exertion to consolidate tranquillity in the far east. In these peaceful aims we signified assent to the proposals of the Japanese government to revise agreements regarding Korean affairs existing between the two governments. However, the negotiations begun upon this subject were not brought to a conclusion, and Japan, without awaiting the receipt of the last responsive proposals of our government, declared the negotiations broken off and diplomatic relations with Russia dissolved.

Without advising us of the fact that the breach of such relations would in itself mean an opening of warlike operations, the Japanese government gave orders to its torpedo boats to suddenly attack our squadron standing in the outer harbor of the fortress of Port Arthur. Upon receiving reports from the viceroy in the far east about this, we immediately commanded him to answer the Japanese challenge with armed force.

Making known this our decision we, with unshaken faith in the Almighty and with a firm expectation of and reliance upon the unanimous willingness of all our loyal subjects to stand with us in defense of the fatherland, ask God's blessing upon our stalwart land and naval forces.

Given at St. Petersburg, January 27, 1904, A. D. (New Calendar, February 9, 1904), and in the tenth year of our reign. Written in full by the hand of His Imperial Majesty. NICHOLAS.

THE BESIEGED CITY.

Port Arthur, as Seen from the North. The City Lies to the East or Left of Tiger's Tail. Portraits: Russian Gen. Stoessel, in command of the defense, at the left; Marshall Oyama, Japanese commander of the besiegers, at the right. This picture covers a broad field, to include the leading fortifications, and details are necessarily reduced.

BRAVE TO THE LAST.

Japanese Deadly Fire from Three Sides Concentrated on a Russian Battery.

CONTENTS

CHAPTER XXVI

GALLANT DEATH OF ADMIRAL MAKAROFF

CHAPTER XXVII

THE JAPANESE MARCH TO THE YALU

CHAPTER XXVIII

THE BATTLE OF THE YALU

CHAPTER XXIX

PORT ARTHUR CUT OFF BY LAND

CHAPTER XXX

DEFEAT OF STAKELBERG'S RELIEF EXPEDITION

ADMIRAL ROJESTVENSKY.
Commanding the Baltic Squadron.

ADMIRAL TOGO.
A Naval Strategist of the First Rank.

GENERAL KUROKI.
One of Japan's Greatest Fighters.

GENERAL OKU.

LIST OF ILLUSTRATIONS

INTRODUCTION

BY TRUMBULL WHITE

THE CAUSES AND SIGNIFICANCE OF THE RUSSO-JAPANESE WAR

EVERY work on an Asiatic subject, whatever its theme or its scope, might fitly begin and end with a query. The whole Asiatic situation, physical, political, social, intellectual and commercial, might be punctuated with interrogation points. It is a continent of questions, a kaleidoscope of problems.

The looming of the war cloud over the Orient raises afresh in the mind of the student of world affairs, the memory of a myriad of problems that have arisen out of Asia, sometimes solved, often given up as too puzzling for human solution. This war between Russia and Japan makes new questions to be asked, that cannot be answered in full until long after the smoke of battle has vanished, and the last echo of cannonading has died away.

We call Africa "the Dark Continent," and the phrase is an apt one to apply to this wilderness of tropical jungle and marsh, of untraveled rivers and ranges, of desert wastes and of benighted peoples who have been left so far behind in the rise of the human race. But Asia is the true Dark Continent, when by darkness we mean mystery, obscurity and uncertainty. It is the birthplace of the human race, of the world religions, of intellectual and social systems of hoary-headed antiquity, in short of Civilization. Yet in spite of all the thousands of years that measure its recorded history there remain vast areas virtually unknown to the world and unpenetrated by even the hardiest explorers, and strange civilizations that we have not begun to understand. Indeed, it is to be doubted if any European or American ever has fully understood the viewpoint, the manner of thought, the ruling spirit of any Asiatic people. Take the wide sweep of Southern Asia that connects Europe with the seat of war in the Orient. Turkey, Persia, Afghanistan, India, Siam, China and Japan. Every one full of mystery when viewed from the standpoint of western civilization; every one known to us in connection with some question. We have had the Eastern Question, the Far Eastern Question, the Indian Question, the Chinese Question, the Korean Question, and now we have the Russo-Japanese Question.

Ever since Peter the Great formed his ambitious plans for Russian aggrandizement two centuries ago, the Asiatic Question has been the Russian Question. As world-interests became more intimate between nations, it has been necessary for the world to ask what Russia is doing. It is a long time since a year has passed without a demand to know what the movements of the great northern power might mean. The world has wanted to know what Russia is doing on the Amur River in Siberia, what in Central Asia, what in Afghanistan, what in Persia, what in Manchuria, what in Korea. At last the demand has become insistent, the problem has become pressing, and Japan has asked the question with a voice that will not be denied answer.

Beside this present cataclysm, such localized political problems as we have had in the Balkans or in Finland become of little moment. It is impossible for the imagination to over-conceive what this grapple between giant foes may mean to the world. Let us look for a moment at the possible results, from the viewpoint of the contestants, taking their positions at their own valuation. The attitude of Russia as voiced by her statesmen and her imperial ruler, is that the world's sympathy should be with their country; that Russia is a Christian nation arrayed against a heathen nation, a white race against a yellow, a western power against an oriental. They declare that a Japanese victory would mean the alliance of Japanese energy, alertness, intelligence, skill and progress with Chinese numbers, resources and endurance for a conquest of all Asia and the virtual commercial dominance of the world. They claim that America and Europe should hope for Russian success, in order that the world shall not be overrun by the Mongols in a movement tremendously greater in its significance than were the invasions of the civilized world a few centuries ago by Ghengis Khan and Tamerlane, or Timur the Tatar.

The Japanese, however, declare that their stand is in the interest of real civilization and true progress, against the onslaught of a tremendous despotism, benighted in its methods of government, oppressive of all liberty and progress, threatening a blight upon freedom wherever its authority extends. Japan claims world sympathy with the clear statement that while Russia may be white of race and Christian of faith, in this instance the Oriental comes far nearer to the ideals of the Christian nations and western civilization than does the power which claims their fellowship.

Prefatory to the body of the volume in hand, let us weigh the merits of these opposing contentions, to learn if it be Russia or Japan that is fighting on the side of occidental civilization. Doing this, we may have some surprises, and even some shocks to preconceived opinion, but that is not as bad as to be resting in an erroneous understanding of conditions. The historical and descriptive chapters which follow will bear out the opinions here summarized, and even the revelations of the war itself, as operations continue and facts develop, will verify the same conclusions, or so it is believed.

First, as to Russian demands for world-sympathy, based on the fact that Russia is a Christian nation, while Japan is pagan. It is, indeed, true that the Russians are among the most religious of peoples, strict in the formal observances required by their faith. The state church is that branch of the Christian religion known as the Greek Catholic, as distinct from the Roman Catholic, the Anglican and the various Protestant churches. In Russia it is all but universally accepted, from prince to peasant, and loyalty to the Emperor, head of the church, exacts loyalty to the church itself. Its cathedrals and its more modest places of worship are everywhere. Its shrines or "ikons," before which prayers are said, hang on the wall of every home, shop and office. But it is difficult to discover that this universal formal orthodoxy has had any effect to uplift the social, mental or moral life of the nation, or, indeed, any good effect of any sort. A low state of morals in private life; a doubtful standard of obligation in public life, commercial and governmental alike; a shocking state of national ignorance; an enslaved thought and speech—these are conspicuous reasons why the real Christianity of Russia may be questioned, however freely we admit the profession of the faith.

In Central Asia, where the Russians have been in authority over the conquered khanates for a quarter of a century, I have looked in vain to find a trace of Christianity, or even the Russian profession of it, given to the native tribes. The Russians have built their own churches everywhere in Turkestan for their own officers, soldiers and merchants, but what we know as missionary effort seems to be absolutely unknown. Furthermore, where Russians rule, missionary effort on the part of other Christian churches is not permitted, and even in St. Petersburg and Moscow, the Protestant and Roman Catholic churches permitted there for the foreign residents exist only by sufferance and must avoid even the suspicion that they are seeking to proselyte.

Let it be clearly understood that Russian dominion over Manchuria and Korea would mean the practical cessation of missionary labors in those fields, and that the Russians themselves would do little or nothing toward a substitute movement of their own church, and does not their chief plea for world sympathy fall to the ground?

It requires but a brief paragraph to show the reverse of this position—the Japanese attitude. True it is that Japan makes no pretense of Christianity. The religion of Buddhism, and certain faiths and practices which we call pagan, are there instead. But Japan makes thought and speech and religion free. There are no trammels on a man's faith, and no restraint upon his teaching.

Where Japanese authority goes, the missionary may labor at will. Liberty and property are protected to a degree unknown in Russia. The spirit of the Island Empire is stirred to freedom, justice, enlightenment, advancement. These are concomitants of Christianity, and the Japanese have truth with them when they declare their nation closer to Christian ideals than is their powerful enemy.

The threat of "The Yellow Peril" which Russia holds up before the world in the demand for occidental sympathy, lacks the aspect of reality. Truly a Japanese-Chinese alliance would be a powerful factor in world affairs, and justly so. But is not the alternative the dominance of northern China by Russia? And should we not rather welcome the uplifting of the Chinese by sympathetic influences, to a position of worth, rather than to approve the suggestion that the ancient nation should either remain an inert mass, or should be dominated and absorbed by jealous European powers for their own profit? The only way in which a Chino-Japanese alliance can invade the world of to-day is by offering the world services or products which are desired. If Japan and China possess qualities which make it possible for them to invade and conquer the world of commerce they will do it in time, in spite of Russian warnings. It is not for Russia to avert that event by a single war, nor can Russia terrify the commercial nations by projecting such a specter against the sky from the rays of the searchlights mounted on their battleships.

It would be impossible to conceive a greater contrast between nations than is immediately apparent between Russia and Japan. In its details the difference will become visible to the most casual reader, on every page of this volume. But we may note here, for the purpose of perspective, some of the larger surface aspects of this contrast. Physically, it is the difference between a continental power nearly three times as large as the whole of the United States without Alaska, and an island power with an area about equal to that of the state of California. The respective populations of the two nations are approximately 135,000,000 and

45,000,000. One is a sparsely settled, little developed land of marvelous natural resources hardly touched as yet, the other is a densely populated land whose productivity is being forced to its highest capacity by the alert, active, thrifty people who dwell there. One sweeps from the Baltic Sea to the Pacific Ocean, and from the burning deserts of Turkestan to the ice-bound shores of the Arctic, including within its limits such tremendous variations of barrenness and fertility, of heat and cold, of plain and forest and mountain as can be paralleled by no single country in the world. The United States and Canada would have to be considered together, in order to find a fair comparison. The other, in its sweep of sea-girt islands reaches from the tropics almost to Behring Sea, but its climate is modified by the great mass of the Pacific, so that except in the extreme north and the extreme south there is a fair degree of uniformity. Generally speaking, Japan is temperate, mild, balmy and beautiful.

In the pages just passed we have drawn some comparisons on the side of government, religion and ruling spirit, and these are yet to be expanded in succeeding chapters. For a comprehensive contrast, the two nations may be fairly typified by a tremendous glacier and a clear mountain torrential stream. One has huge mass, slow motion, great weight, resistless progress toward the warmer valley, and at last the sea. Cold, silent, crushing, it makes its way over obstacles to its destination. What cannot move out of its path it destroys. On its surface appears the debris gathered in its frozen flood. The other moves in haste, clears away obstacles by its impact, leaps down the slope of beetling cliffs, turns mill wheels for men, gives water to thirsty fields in the valleys, lends beauty and cheer to the landscape, and in the end reaches the sea to rest after worthy toil well done.

If in the end we are asked to say which has the greater power, we can but point to the huge gorges and canons cut through mountain ranges by brooks hardly wider than a man might leap, to show the amazing effects of erosion. May we not yet find Japan making deeper impress on world affairs than can Russia, with all her glacial momentum?

Most wars are but episodes in history, a partisan controversy, a sovereign's ambition, a greed for plunder, as the stimulating cause, with indifferent importance to the world in the outcome. But this Russo-Chinese conflict looms big in its ultimate consequences. It promises to affect the destinies of the world, to decide who shall dominate the Orient, to leave its traces on the commerce, the politics and the industries of nations on the other side of the world who have no share in the fighting.

If there be one great naval engagement or one great land battle which, when the war is ended, can be named as the turning point in the conflict, that struggle will go down into history as one of the decisive battles of the world. Those of like significance are few, indeed. Charles Martel defeated the Saracens and it was determined that Europe was to be Christian instead of Moslem. The destruction of the Spanish Armada by British ships and British storms decided that England, rather than Spain, was to become the dominant power in Europe, and that Anglo-Saxon, rather than Spanish, ideals were to survive. Marathon, Saratoga, Waterloo, Gettysburg, perhaps a dozen other battles, might be named as really decisive in world history, and to these some great name will be added before the Russo-Japanese war is ended.

We of America are specially concerned in the present war. A Pacific Ocean power, as the United States has become, we are, perhaps, more vitally interested in the outcome of this war, and the ultimate results of the outcome, than in any foreign war that has been waged heretofore. Our states face westward to the Orient, with great cities and busy harbors, fertile fields and noisy factories to be reckoned in world affairs. From Alaska, on the north, to Samoa, on the south, and from San Francisco to Manila lie the American possessions. Not only Hawaii, but Tutuila, Guam, the Philippines and many a coral islet yield to American sovereignty. The only trans-Pacific cable is American. Our ships ply every waterway on the great ocean. Our merchants are in every port. Our Chinese neighbors are the most vital problem in the labor conditions of the Pacific states and the island dependencies. The possible restraint of trade with the Orient in the event of Russian victory is a real factor in the situation, as it affects our commercial relations with China.

This war then is to be far-reaching in its consequences, and it is to settle many things. I quote from a recent expression which summarizes the situation. "The question in brief is: Shall the reactionary principles of government, education and religion, represented by the Russian autocracy, be extended over southern Asia as well as over northern, or shall the spirit of occidental civilization prevail? That occidental civilization is suitable and adaptable to the needs of oriental people, and that it is of unspeakable service to them, is demonstrated by the case of Japan, which, in a comparatively short time, has taken rank among the civilized powers of the world. It is a remarkable thing that in this war the interests of a true Christian civilization, the broad tolerance which permits everyone to exercise his religious faith and political opinion, whatever it may be, without interference, are championed by a quasi-pagan nation, while intolerance, oppression and prohibition of freedom of thought, religious or political, characterize its quasi-Christian opponent."

CHAPTER I

THE WAR AND ITS CAUSES

Japan and Russia Clash in Deadly Conflict for Supremacy in Asia—News Causes a Sensation in the United States and Japan's Victory is Likened to Dewey's in Manila Harbor—Sympathy of American People for the "Under Dog"—Both Nations Issue Formal Statements Concerning the Causes of the War.

AT midnight, February 8, 1904, a flotilla of Japanese torpedo boats stole into the roadstead at the Russian stronghold of Port Arthur, blew up two of the finest battleships and one of the fastest cruisers in the Russian navy and escaped without a scratch.

Such was the news that electrified the world the following morning, and told, although no declaration of war has been issued by either power, that Japan and Russia at last had grappled with each other in the long-expected struggle for supremacy in Asia.

This news created a sensation in the United States almost as great as Dewey's May Day victory in Manila harbor, and expressions of admiration were heard on every hand for the brave little islanders who had dared to clip the claws of the great Russian bear and defend their nation against his further encroachments.

A very large per cent of Americans, especially those who had followed the details of the quarrel between the two nations, were in full sympathy with the Japanese, and justified Japan in striking the first blow as the only means of preventing Russia from massing a large army in Manchuria and Korea and overpowering her by force of numbers. The situation was similar to the opening of hostilities in the Boer-British war, when Great Britain delayed negotiations while she poured troops into the Transvaal and thus compelled the Boers to strike first to save themselves from being overwhelmed.

American Sympathy for "the Under Dog."

Americans naturally sympathize with the "under dog," and the "little fellow," hence the spectacle of that little nation of little brown men striking the powerful Russian such a terrible blow, stirred the red blood in their veins and aroused a feeling that almost could be called enthusiasm.

This war which began with such dramatic effect was not unexpected, although a month before the Whitehead torpedoes tore their way into the vitals of the three great warships in the roadstead at Port Arthur, there was reason for belief that Russia would accede to the demands of Japan and that a terrible conflict at arms would be avoided. Almost before the hope of peace had found lodgment in the minds of the people, the war cloud which had been hovering for so long over that part of Asia burst with a lightning flash.

In the introduction to this work Mr. Trumbull White, who has traveled all over Russia and Japan and is as familiar with the characters of the Muscovite and Malay as he is with the topography of their respective countries, has set forth the fundamental differences of the two races and has stated the chief causes of the war which has been so long in brewing.

These differences and causes did not reach an acute stage until Russia violated her treaty obligation to evacuate Manchuria on the date set. Even this failure to keep faith with the powers at interest might have been condoned had it not been for the constant encroachment upon Korea, which, had it been allowed to proceed, would have brought Russian influence to the very door of Japan. Then it was that Japan through her foreign office began to make protests and demands. This correspondence was kept secret until the crisis was reached on January 6, when Baron de Rosen, the Russian Minister to Japan, was ordered to leave Tokio, and Mr. Kurino, the Japanese Minister to Russia, was told to demand his passports.

Then the following statements concerning the correspondence which had passed between the governments was given to the public.

Japan's Statement.

The statement issued by Japan is as follows:

"Section 1. It being indispensable to the welfare and safety of Japan to maintain the independence and territorial integrity of Korea and to safeguard her paramount interests therein, the Japanese government finds it impossible to view with indifference any action endangering the position of Korea, whereas Russia, notwithstanding her solemn treaty with China and her repeated assurances to the Powers, not only continues her occupation of Manchuria, but has taken aggressive measures in Korean territory.

"Should Manchuria be annexed to Russia the independence of Korea would naturally be impossible. The Japanese government therefore being desirous of securing permanent peace for eastern Asia by means of direct negotiations with Russia with the view of arriving at a friendly adjustment of their mutual interests in both Manchuria and Korea where their interests meet, communicated toward the end of July last such desire to the Russian government and invited its adherence. To this the Russian government expressed a willing assent.

What Japan Proposed.

"Accordingly on August 12 the Japanese government proposed to Russia through its representatives at St. Petersburg the basis of an agreement which was substantially as follows:

"1. A mutual engagement to respect the independence and territorial integrity of the Chinese and Korean empires.

"2. A mutual engagement to maintain the principle of an equal opportunity for the commercial industry of all nations with the natives of those countries.

"3. A reciprocal recognition of Japan's preponderating interests in Korea and that Russia has special interest in railway enterprises in Manchuria, and a mutual recognition of the respective rights of Japan and Russia to take measures necessary for the protection of their above-named interests.

"4. The recognition by Russia of the exclusive rights of Japan to give advice and assistance to Korea in the interest of reform and good government.

"5. The engagement on the part of Russia not to impede the eventual extension of the Korean railway into southern Manchuria so as to connect with eastern China and the Shanghaikwan-Newchwang lines.

Russia Charged with Delay.

"It was the intention of the Japanese government originally that a conference should take place between the representatives at St. Petersburg and the Russian authorities, so as to facilitate progress as much as possible in reaching a solution of the situation, but the Russian government absolutely refused to do so, on the plea that the Czar planned a trip abroad, and for other reasons it was unavoidably decided to conduct the negotiations at Tokio. It was not until October 3 that he Russian government presented counter-proposals and in them she declined to engage in respect to the sovereignty and territorial integrity of China and stipulated the maintenance of the principle of equal opportunities for the commerce and industry of all nations in China, and requested that Japan declare Manchuria and its littoral as being entirely outside of her sphere and interest.

"She further put several restrictions upon Japan's freedom of action in Korea. For instance, while recognizing Japan's right to dispatch troops when necessary for the protection of her interests in Korea, Russia refused to allow her to use any portion of Korean territory for strategical purposes. In fact, Russia went so far as to propose to establish a neutral zone in Korean territory north of the thirty-ninth parallel. The Japanese government utterly failed to see why Russia, who professed no intention of absorbing Manchuria, should be disinclined to insert in the convention a clause in complete harmony with her own repeatedly declared principle respecting the sovereignty and territorial integrity of China.

Japan's Interests in Manchuria.

"Furthermore, the refusal of the Russian government impressed the Japanese government all the more with the necessity of the insertion of a clause to the effect that Japan has important commercial interests in Manchuria and entertains no small hopes of their further development, and that politically Japan has even interests there by reason of its relations to Korea, so that she could not possibly recognize Manchuria as being entirely outside her sphere of interests. These reasons decided Japan to absolutely reject the Russian proposal. The Japanese government explained the above views and at the same time introduced other necessary amendments in the Russian counter-proposal. They further proposed with regard to a neutral zone that if one was to be created it should be established on both sides of the boundary line between Manchuria and Korea with equal width, say fifty kilometers.

Clauses on Manchuria Suppressed.

"After repeated discussions at Tokio the Japanese government finally presented the Russian government its definite amendment on October 13. The Japanese government then frequently urged the Russian government for a reply. In that reply Russia suppressed clauses relating to Manchuria so as to make the proposed convention apply entirely to Korea and maintained its original demand in regard to the non-employment of Korean territory for strategical purposes, as well as a neutral zone, but the exclusion of Manchuria from the proposed convention being contrary to the original object of the negotiations, which was to remove causes of conflict between the countries by a friendly arrangement of their interests both in Manchuria and Korea, the Japanese government asked the Russian government to reconsider the question and again proposed the removal of the restriction regarding the use of Korean territory and the entire suppression of the neutral zone on the ground that if Russia was opposed to the establishment of one in Manchuria it should not establish one in Korea.

Russia's Proposal Unsatisfactory.

"The last reply of Russia was received at Tokio on February 6. In this reply it is true Russia proposed to agree to insert the following clause in the proposed agreement: The recognition by Japan of Manchuria and its littoral as outside her sphere and interests, whilst Russia within the limits of that province would not impede Japan or any other powers in the enjoyment of rights and privileges acquired by them under existing treaties with China exclusive of the establishment of settlement, but this was proposed to be agreed upon only upon conditions maintaining the clauses regarding a neutral zone in Korean territory and the non-employment of Korean territory for strategical purposes, already fully explained to them. It should further be observed that no mention was made at all of the territorial integrity of China in Manchuria, and it must be self-evident to everybody that the engagement now proposed by Russia would be unpractical in value, so long as it was unaccompanied by a definite stipulation regarding the territorial integrity of China in Manchuria, since treaty rights are only coexistent with sovereignty.

"Eventually absorption of Manchuria by Russia would annul at once those rights and privileges acquired by the powers in Manchuria by virtue of treaties with China."

Russia's Side of the Story.

The Russian account of the negotiations which led to the rupture is as follows:

"Last year," says the foreign-office statement, "the Tokio cabinet, under the pretext of establishing the balance of power and

"WORN OUT."
Dawn at Port Arthur, the Besieged City.

THE DESPERATE FIGHT.

In the Sea at Nanshan. The Flying Russians Slaughtered.

a more settled order of things on the shores of the Pacific, submitted to the imperial government a proposal for a revision of the existing treaties with Korea.

"Russia consented and Viceroy Alexieff was charged to draw up a project for a new understanding with Japan in co-operation with the Russian minister at Tokio, who was intrusted with the negotiations with the Japanese government. Although the exchange of views with the Tokio cabinet on this subject were of a friendly character, Japanese social circles and the local and foreign press attempted in every way to produce a warlike ferment among the Japanese and to drive the government into an armed conflict with Russia. Under the influence thereof the Tokio cabinet began to formulate greater and greater demands in the negotiations, at the same time taking most extensive measures to make the country ready for war.

Takes Military and Naval Measures.

"All these circumstances could not, of course, disturb Russia's equanimity, but they induced her to also take military and naval measures. Nevertheless, to preserve peace in the Far East, Russia, so far as her incontestable rights and interests permitted, gave the necessary attention to the demands of the Tokio cabinet and declared herself ready to recognize Japan's privileged commercial and economic position in the Korean peninsula, with the concession of the right to protect it by military force in the event of disturbances in that country.

Points Demanded by Russia.

"At the same time, while rigorously observing the fundamental principle of her policy regarding Korea, whose independence and integrity were guaranteed by previous understandings with Japan and by treaties with other powers, Russia insisted on three points:

"1. On a mutual and conditional guarantee of this principle.

"2. On an understanding to use no part of Korea for strategic purposes, as the authorization of such action on the part of any foreign power was directly opposed to the principle of the independence of Korea.

"3. On the preservation of the full freedom of navigation in the straits of Korea.

"The project elaborated in this sense did not satisfy the Japanese government, which in its last proposals not only declined to accept the conditions which appeared as the guaranty of the independence of Korea, but also began at the same time to insist on provisions to be incorporated in a project regarding the question of Manchuria. Such demands on the part of Japan, naturally, were inadmissible, the question of Russia's position in Manchuria, concerning in the first place China, but also all the powers having commercial interests in China.

Drew Line at Manchuria.

"The imperial government, therefore, saw absolutely no reason to include in a special treaty with Japan regarding Korean affairs, any provisions concerning territory occupied by Russian troops. The imperial government, however, did not refuse, so long as the occupation of Manchuria lasts, to recognize both the sovereignty of the Emperor of China in Manchuria and also the rights acquired there by other powers through treaties with China. A declaration to this effect has also already been made to the foreign cabinets. In view of this the imperial government, after charging its representative at Tokio to present its reply to the last proposal of Japan, was justified in expecting the Tokio cabinet to take into account the considerations set forth above and that it would appreciate the wish manifested by Russia to come to a peaceful understanding with Japan.

"Instead of this the Japanese government, not even awaiting this reply, decided to break off negotiations and to suspend diplomatic relations. The imperial government, while laying on Japan the full responsibility for any consequences of such a course of action, will await the development of events and the moment it becomes necessary will take the most decisive measures for the protection of its rights and interests in the Far East."

The events which led to the diplomatic correspondence referred to in the foregoing official statements cover only a period of eight years.

It was in 1896 that China was first led to grant certain railway-making facilities to Russia in Manchuria. She was said to have then resolved, in recognition of Russia's friendly attitude toward her at the time that she was at war with Japan, to grant to Russia the right to build and control certain railways passing east and west through Manchuria. Russia also acquired the right, in case of war, of using Port Arthur as a base for military operations. Russia, moreover, on the plea that the needful control of the railway could not otherwise be assured, was to be permitted to some extent to employ her military forces in that region.

March 27, 1898, Russia was granted by China a twenty-five years' lease of Port Arthur and the adjacent bay of Talien-Wan. Russia likewise acquired the right of building railways to the same points. Russia promised that both ports should be open to foreign trade and to the ships of all friendly nations. How lightly that promise weighed on the Russians has since been manifested. The convention of April 8, 1902, between Russia and China, however, provided for the evacuation of Manchuria by Russia under certain conditions, beginning with the province of Mukden (i. e., Shing-King, or southern Manchuria). The Kirin province (central Manchuria) was to be evacuated October 8, 1902, and the Hei-Lung-Chuang province (northern Manchuria) April 8, 1903. The promise was not kept, but April 18, 1902, Russia presented a series of demands on China, which were to be complied with prior to the evacuation being begun (already ten days late). These demands covered exclusive political and commercial rights for Russia throughout Manchuria. China refused. The United States, which has been directly promised by Russia that the open-door principle should be maintained in Manchuria, was at the time asking China, and so was Japan, that certain places in Manchuria should be opened to foreign trade. Russia held out for a time, but consented in July, 1903.

The treaties referred to were signed between China and Japan and the United States respectively in October, 1903, and were ratified, in spite of Russia's vigorous protests, January, 1904.

The Russian minister at Pekin further engaged, September 6, 1903, that Mukden province should be evacuated October 8, but the engagement was accompanied by conditions impossible of acceptance, and after a pretense of the removal of troops had been made the force was reinstalled and the city of Mukden itself, sacred to the Chinese as the birthplace of the present Manchu dynasty, was ostentatiously reoccupied. Such is the history of Russia's dealings with respect to Manchuria. Since the fall of 1903 there have been frequent exchanges of negotiations between Japan and Russia, and Japan, feeling that Russia had made no effort to meet its terms, forced the issue by war.

CHAPTER II

THE FIRST BLOW

Japanese Torpedo Flotilla Attacks the Russian Fleet in the Roadstead at Port Arthur and Blows Up Two Battleships and a Cruiser—Japanese Lose Two Torpedo Boats—Japanese Battlefleet Returns the Following Morning and Engages the Russian Ships and the Land Battery in Fierce Action.

THE beginning of the Russo-Japanese war was marked by an incident similar to the close of the Napoleonic wars, but with a different termination. The night before the battle of Waterloo the British officers were at a dance, celebrated in Byron's immortal lines:

> "There was a sound of revelry by night
> And Belgium's capital had gathered there."

As the first gray streaks of dawn dimmed the luster of the ball room the boom of Napoleon's heavy artillery was heard, and the scene of revelry was changed to one of consternation.

On the night when Japan struck the first blow the Russian vice-regal naval officers in command of the big squadron of warships in the harbor of Port Arthur were at a reception when they heard three explosions that shook the town, followed by the roaring of the big guns of the battery on the hill commanding the harbor and the rapid fire of the guns on their own vessels.

It was about midnight when the officers and men left on watch on the Russian warships saw six small craft approaching, but the watchers were thrown completely off their guard by the fact that the approaching vessels showed lights, funnels and signals exactly like those on the Russian ships. Nearer and nearer crept the six little engines of destruction until they were within half a mile of the big ships of the Russian squadron and then each of them discharged a torpedo and fled.

Three of the deadly missiles went straight to their marks and three terrific explosions followed. The monster battleships Czarevitch and Retzivan lurched and listed as if about to keel over and the magnificent swift cruiser Pallada began to sink. Wounded to the death they turned and made for the harbor entrance and ran upon the beach to escape sinking.

A continuous fire from the ships and forts was poured upon the Japanese torpedo boats, four of which escaped. One, however, was sunk, and another, in a sinking condition, was deserted by the crew and was afterward captured by the Russians.

Then the Russians saw several Japanese cruisers in the distance. The remainder of the fleet was still farther off, lying to the eastward and showing searchlights.

After the retirement of the Japanese torpedo boats the Russian steamer squadron, under Admiral Prince Moktompsky, followed to investigate and then returned. The action ceased at 3 a. m.

Russian Losses in Men.

The loss on the Russian ships was eight killed and twenty wounded. Apart from the disablement of the three Russian ships the damage done to the fleet and forts was not very great. There were many Russian torpedo boats and destroyers in the harbor, but they were not ready to resist the attack of the Japanese, who, in fact, created a great deal of surprise, not only by their unexpected onslaught, but by the promptness and bravery with which they acted.

The Czarevitch, which, with the Retvizan and Pallada, was disabled by the Japanese torpedo boats, was a battleship of 13,110 tons, built in France. In armor, armament and speed she about equals the Retvizan.

The battleship Retvizan was built by the Cramps at Philadelphia. She was of 12,700 tons displacement, had 16,000 indicated horse-power, and had a speed of eighteen knots per hour. Her armor was of Krupp steel, from four to ten inches in thickness, and her armament consisted of four 12-inch guns, twelve 6-inch guns, twenty 3-inch guns and twenty-six smaller rapid-fire guns.

The Pallada was a cruiser of 5,630 tons. She was built in Germany and was completed in 1902. Her armament consisted of six 6-inch guns, twenty 3-inch guns and eight 1.4-inch guns. Her speed was estimated at twenty knots.

The following morning the news arrived at Port Arthur from Dalny that the Japanese fleet was steering westward in attack formation. It came in sight about 11 o'clock. There were in all fifteen ships in two lines of battle—six battleships, six first-class cruisers and three second-class cruisers. The Russians had outside thirteen large vessels under Admiral Stark on the flagship Petropavlovsk and Rear Admiral Prince Moktompsky on the flagship Peresviet, excluding the Pallada and Czarevitch, flagship of Rear Admiral Mollas, and the Retvizan, which was lying aground across the inner harbor entrance, and it was low water.

Action was commenced by the guns of the land battery. The morning broke dull, with a light wind, and the heavy smoke rendered it difficult to observe the details of the action, but those who saw the battle witnessed all that was possible to be seen from Beacon Hill, opposite the entrance to the harbor, and in the line of fire. Two shells fell near the spectators and about twenty others fell in the old town and the western harbor, where many steamers flying neutral flags were anchored. After the commencement of action all the people fled toward the hill outside the town for protection.

A little while after the first spell a big twelve-inch gun exploded, smashing the office fronts of the Guensberg Yalu Concessions Company and the Russo-Chinese bank. The streets were then entirely deserted, but the local police kept splendid order and there was no looting. The women and children were very brave.

Japanese Battleship Destroyed.

Regiments from adjoining barracks and camps came pouring through town to take up defensive positions in the event of the

Japanese landing. The Japanese warships steamed slowly past in line of battle to the westward and about four miles off, each vessel beginning to fire when opposite the Russian ships, which were two miles off shore.

The action became general. There was no maneuvering, simply heavy and fast firing on both sides. One war correspondent counted over three hundred shells, few of which reached their mark, and the others did not explode.

During the action several merchant steamers outside the roads moved their position, but none was allowed to leave its anchorage in the harbor. The firing ceased at noon, the Japanese ships withdrawing to the southward, having lost one battleship, and one large cruiser was put out of action. One small boat was chased and sunk by the Novik, which afterward received a shell at the water line, but reached port all right, Admiral Stark signaling "Well done!" while all the rest of the fleet cheered her arrival.

Even the three ships aground fired during the action. Afterward the Czarevitch got off at high water and was towed into a large basin for repairs. The Pallada effected her own repairs and rejoined the fleet. The Retvizan remained aground.

The casualties were twenty-two killed and sixty-four wounded. Nearly half the casualties occurred on the Pallada and Novik. The Japanese fleet sailed southward and at 1 o'clock all was quiet. The wounded were taken ashore and removed to hospitals.

After Monday night's action many Japanese torpedoes were found floating outside the harbor. They were secured and their mechanism extracted. During the afternoon Viceroy Alexieff ordered all the women and children and non-combatants to leave and the slow special trains were crowded. They ran as often as possible to and from Dalny. The women and children were immediately removed in an English steamer.

After the action on Monday night official telegrams from Vladivostok stated that the cruiser squadron, consisting of the Cromoboi, Rossia, Rutik and Bogatyr, had shelled a town in Yezo and then returned to Vladivostok.

THE CZAR ADDRESSING A BODY OF IMPERIAL GUARDS.

Nicholas II, the Czar of all the Russias, who is said to have wept bitterly when the news of war was brought to him, quickly rallied from his depression, and issued a series of manifestoes to his army and navy calling for vengeance upon the enemy.

A SIGNAL TORCH AT A COSSACK POST.

IN THE scouting operations of the Cossacks in Manchuria, the country was so broken and rugged that it was often difficult for scattered bodies to keep in communication. At times they were obliged to fall back upon their ancient way of signaling by means of huge torches.

RUSSIANS DRAWING LOTS FOR MILITARY SERVICE.

Russia's almost incalculable military strength made it unnecessary that all men fit for military duty should serve. The recruits were therefore required to draw lots. This scene is in a Russian town hall at the opening of the war.

THE EMPEROR OF JAPAN AND HIS SUITE.

The stern yet calm face of Japan's ruler is well set forth in the above. It is the face of a warrior and a thinker. In the background are typical Japanese officers grouped around the imperial standard.

The Russian Version of the Battle.

Viceroy Alexieff's official report of the naval engagement on February 9 is as follows:

"After the night torpedo attack the Japanese fleet, consisting of sixteen war ships, appeared at 10 o'clock in the morning off Port Arthur. Its appearance was noted by the coast signal stations, as well as by the ships of our squadron, which lay in the outer roadstead, fully prepared for battle. Our squadron consisted of five battleships, five first and second-class cruisers and fifteen torpedo boats, under the command of Vice Admiral Stark and Rear Admiral Ukhtonski. The coast batteries immediately prepared to receive the enemy. Our squadron weighed anchor in order of battle, and, upon the first shot being fired by the enemy, the fleet and batteries replied simultaneously with a lively cannonade. The most exposed to the enemy's fire were the ships of the squadron, battery No. 15, on Electric cliff, and battery No. 13, on Golden Hill. Other coast batteries, principally numbers 17 and 18, were also under fire.

"The following vessels took part in the battle:

"In the front line were the battleship Petropavlosk, flying the flag of Vice Admiral Stark and commanded by Captain Jakovloff; the battleship Sevastopol (Captain Chernychoff), the battleship Poltava (Captain Usphenski), the battleship Peresviet, flying the flag of the Rear Admiral Ukhtonski, commanded by Captain Boissmann, and the battleship Pobieda (Captain Sazareuny). There were also the first-class cruisers Bayan (Captain Viron), Diana (Captain Satouski) and Askold (Captain Gramchikoff), the second-class cruisers Boyarin (Captain Sarycheff) and Novik (Captain Jassen).

"The battleships Czarevitch (Captain Gregorovitch) and Retvizan (Captain Sensnovitch) and the cruiser Pallada (Captain Kossovitch), which lay at anchor, having been damaged in the torpedo attack of the night before, likewise participated in the fight, as did the transport Angara (Captain Suchonlin). There were also engaged the torpedo boats of the first and second divisions, under Divisional Commanders Matussevitch and Duenter, as follows: Vnimatelni (Captain Simon), Vlastini (Lieutenant Karzeff), Voevoi (Captain Yelisseitff), Rezstrashni (Captain Zimmermann), Resposhtchadni (Lieutenant Lukin), Vnushitelni (Lieutenant . Povushkinp). Vuinoslivi (Lieutenant Reichter), Grozovoi (Lieutenant Scheltinga), Razyaschtchi (Captain Simonoff), Ryoshitelni (Lieutenant Kornilieff), Silny (Lieutenant Kodorovitch), Stereguschtchi (Lieutenant Kusmenkaravayeff), Storczhevei (Captain Kilkin), Smyeli (Captain Schutz) and Serditi (Lieutenant Kusmenkaravayeff, Jr.).

Lays to Right of Squadron.

"Throughout the fight the torpedo division lay to the right of the squadron, a distance of from ten to fifteen cabeltari (a Russian sea measure), awaiting orders.

"The land batteries were under the general direction of Major General Baloff, commander of the Kwan-Tung garrison of artillery. Several vessels of the enemy's fleet sustained damage, which explains why they avoided further fighting, although they were much superior to us in strength.

"According to the reports of the commanders the men fought exceedingly well, on which account in virtue of the imperial authorization I have conferred six crosses of the Order of St. George on each company of the first and second class warships having crews of over 200 men, four crosses on each company of all other ships of the second class, one cross on each torpedo boat, one cross on the signal station at Golden Hill, which operated under the heavy fire of the enemy; four crosses on battery No. 15, three on battery No. 13, and one cross on a gunner, who, though severely wounded, returned to his battery.

"Our losses were:

"Of the squadron—Five officers wounded, fourteen men killed and sixty-nine men wounded.

"Of the fortress and garrison—One man killed, one man severely wounded and five men slightly wounded.

"In announcing the above to your majesty I am pleased to be able to add that the naval and land forces in the far East are inspired by the most heartfelt wish to meet their insolent enemies breast to breast, in order to fulfill their duty in sight of their adored ruler, and, firm and unshakable, to fight for the honor and glory of their beloved fatherland."

When Admiral Togo resolved upon his midnight torpedo attack, so characteristic of Japanese methods and courage, his squadron was fifty or more miles from Port Arthur, and, of course, the attempt to enter or approach the port was attended with risks of the gravest description. The Japanese, however, were in no way dismayed, and, by dint of very clever tactics, they succeeded in eluding the Russian torpedo boats on picket duty and getting near enough to the fleet of battleships and cruisers, lying at anchor under protection of the guns of the forts, to discern the men moving about on their decks. The flotilla consisted of the destroyers Asashio, Shirakumo, Akatsuki, Kazumi, Ikadzuchi, Oboro, Inadzuma, Usugomo, Shinonome and Sasaname, ten in all.

The vessels immediately set about discovering the location of the Russian fleet, and, although several of the enemy's picket boats were on the watch, they succeeded by brilliant strategy in getting within striking distance. This was in no small measure due to the excellence of the Japanese intelligence service, as is evident from the fact that practically all the Russian vessels were found to be in the precise positions where Captain Asai, who was in command of the flotilla, had been told he would probably find them.

Operated in Three Sections.

To insure effective work over a wide area the flotilla was divided into three sections, Captain Asai having the Asashio, Shirakumo, Akatsuki and Kasumi; Lieutenant Ishii the Ikadzuchi, Oboro and Inadzuma, and Lieutenant Tsachiya the Usugomo,

Shinonome and Sasanami. The intrepid leaders ran in as close as possible to the enemy and deliberately selected the largest of the Russian warships as the objects of their attack. The audacity of the movement was rewarded by complete success, for the Russians were taken entirely by surprise, and when the attack commenced were thrown into a condition bordering on panic.

This much can be gathered from the circumstances, that, although the Russian fleet opened a heavy fire and got its searchlights to bear, the Japanese destroyers were able to discharge their torpedoes and escape with practically no damage to themselves. One destroyer—the Inadzuma—missed fire, and her plucky commander coolly turned her round and fired the torpedo again. Each daring little vessel discharged two torpedoes and then steamed out to rejoin the fleet, keeping well in shore until out of range of the guns of the forts. When dawn broke three of the finest ships of the Russians—the battleships Czarevitch and Retvisan and the armored cruiser Pallada—were either in a sinking condition or on shore. The Pallada was run on shore in order to save her from going down, and both the battleships had to be taken into the inner harbor into shallow water to prevent them from sinking.

Exploit Was Brilliant.

There can be no doubt that the whole affair constitutes a naval exploit of the most brilliant description. To creep into a hostile anchorage, eluding the enemy's vessels on the lookout for an attack, and effectually to cripple three of the most important of the enemy's ships, is a feat of a very notable character, and it puts the seal on the reputation in naval matters which Japan gained in the much less formidable matter of the war with China.

This successful torpedo destroyer engagement was followed up on the next morning by a general attack by the Japanese fleet. Admiral Togo endeavored to entice his antagonists into a fight in the open sea, but failing in this, he passed Round island on the right and proceeded in a single line toward Port Arthur, leading in the battleship Mikasa. Each of his ships opened fire on passing the enemy's fleet, most of which remained at anchor under the guns of the forts. The Russian ships responded vigorously, and for some forty minutes a heavy exchange of firing took place.

It is admitted by the Japanese that several of the Russian ships fought extremely well, particularly the Novik, which did very good work, but Admiral Togo's ships more than held their own and inflicted severe damage on their opponents.

The Askold afterward sunk in the harbor, and three other cruisers stood in urgent need of repairs. The Bayan and a ship of the Poltava type were among those seriously injured. The remarkable feature of the fight was the accuracy of the Japanese fire, notwithstanding the fact that the range was 8,000 yards. Scarcely less notable was the immunity of the Japanese vessels, which sustained surprisingly little damage. The Iwate (armored cruiser) was hit by a 10-inch shot from the Novik and the Fuji (battleship) was struck on her forward funnel; but otherwise very little injury was sustained by the courageous islanders. There were, however, a considerable number of men killed and wounded, and with regard to these several striking stories are told, apparently on good foundation.

The most ghastly is that relating to Lieutenant Miura, who was struck by a shell while standing on the bridge of the Fuji. A piece of his belt was the only relic which was left of the unfortunate officer. Less grewsome and more agreeable are some of the incidents evincing the indomitable spirit which animates the Japanese engaged in active service. A midshipman on the Fuji was mortally wounded. He was at once conveyed below and surgically attended to, and on the attendant's endeavoring to remove his shoes he protested loudly, declaring that he was going back to the fight at once. His next question was, where was his missing limb. "I shall be able to go on deck again in a few minutes!" he pathetically exclaimed, but a very short time afterward his patriotic ardor was quenched in death.

Story of a Russian Naval Officer.

An officer on the cruiser Pallada furnished an interesting account of experiences aboard his warship on the night of the torpedo attack.

The captain of the Pallada, which vessel occupied the advance position, had descended from the bridge for a last look around before going below when he perceived lights advancing. The ships' lights shown were white above red, being those of the Russian warships when entering the harbor. The captain of the Pallada supposed the approaching vessels were the Russian torpedo boats returning from Dalny, and his suspicions were only aroused when upon drawing nearer they covered and uncovered their lights at irregular intervals. The signal tower signaled that the lights were not understood.

At this moment the Pallada's captain, through the thick night, made out the outlines of the torpedo boat destroyers' smokestacks in pairs, amidships. As the stacks of the Russian destroyers are in line fore and aft, the crews of the Russian ships were instantly called to quarters.

In less than three minutes the Pallada's men were at their posts, orders were given that the guns be charged with grape, and a fierce fire was opened on the oncoming Japanese. The batleships Czarevitch and Retvizan, which were in the first line, a short distance astern of the Pallada, also opened fire almost immediately.

A terrific explosion occurred under the hull of the Pallada, raising a torrent of water which submerged the cruiser's deck, but did not stop its firing or maneuvering, which now, however, were complicated by the measures taken to close a breach amidships, below the waterline, made by the explosion of the torpedo.

Soon afterwards two other torpedoes exploded almost simultaneously, one under the bow of the Retvizan and the other under the stern of the Czarevitch. This double explosion ended the attack, the Japanese vessels retiring at full speed.

CHAPTER III

THE RUSSIAN ADVANCE ACROSS ASIA

BY TRUMBULL WHITE.

Historical and Geographical Sketch of the Great Northern Power—Peter the Great and His Boundless Ambitions—Yermak, the Conqueror of Siberia—Muraviev Carries Russian Dominion to the Pacific Ocean—The Great Siberian Railway—Manchuria in the Grasp of the Muscovites—The Conquest of Turkestan—Resources and Industries of the Russian Empire—The People and their Customs—The Siberian Exile System.

STUDENTS of history, seeking the significant facts in the life of any nation, find that there is no more illuminating method of procedure than to select the successive great figures in the nation's progress, and, by studying these individuals, learn of the country of which they were a part.

Russia is one of the best exemplars of this biographical method of history study. Despotism and autocracy as it is, the great men and women of Russia have embodied the nation's history, while the people in the mass have been little more than raw material with which the ones in power have worked their purpose. With the increase of enlightenment and liberty, the accuracy of this statement always will tend to become less, but thus far in the history of the Russian Empire we may take it literally.

Not always are these conspicuous figures the rulers of the land. Sometimes they are the talented servitors of dullard sovereigns, sometimes progressive rebels against unenlightened authority, intellectual, physical or otherwise. But, however that may be, it is they who make the history, and in holding them up to view we obtain the graphic quality to a degree that never accompanies impersonal history.

Rurik, Vladimir, Ivan the Terrible, Peter the Great, Catherine, Alexander, Nicholas—these are some of the sovereigns to be observed in any historical sketch of Russia. Yermak, Muraviev, Khabarov, Skobelev, Cassini, Pobyedonostsev, Witte, Khilkov, Alexiev, Kouropatkin—these are some of those who have served their country, past and present, as warriors, statesmen or diplomates, and who must be noted now. If to these we add the familiar names of others now in the service of imperial Russia, we have the essentials of a historical sketch such as will serve our purpose here.

Before we attempt to view Russians singly, let us see whence Russians came and where they dwell.

Right across the northern half of the two continents, Asia and Europe, sweeps an almost continuous plain, from the shores of the Baltic and Black Seas to the Pacific Ocean. The mountain ranges which traverse this plain are not such as to form stern obstacles for wandering tribes, and this physical fact has been of great importance in affecting the political and industrial conditions. The primitive nomadic races of the past were able to wander virtually from sea to sea, and in the more recent historical periods there were no permanent obstacles to the sweeping conquest of a Ghengis Khan from Asia, nor to the extension of authority of the Russian conquerors from Europe. So the tide of conquest of late years has but reversed itself, and the European horde is repaying to the Asiatics the debt of past centuries.

We know that the civilization of southern Europe fell a prey to the incursions of Asiatic tribes, and of Germanic tribes pressed southward by Asiatic invaders. The Goths, the Vandals, the Huns, the Mongols, Attila, Alaric, Genseric—these are names familiar enough in our reading of general history. But there are others as terrible and as familiar in Russian history, the names of races and chieftains who swept into northern Europe from Asia, hardly known to any but specialists in history, so little influence had they upon our immediate progenitors.

The Russian is of that racial group which we call the Slav. Our earliest knowledge of the race dates but to the ninth century, when they were dwellers in the region now included in Prussia,

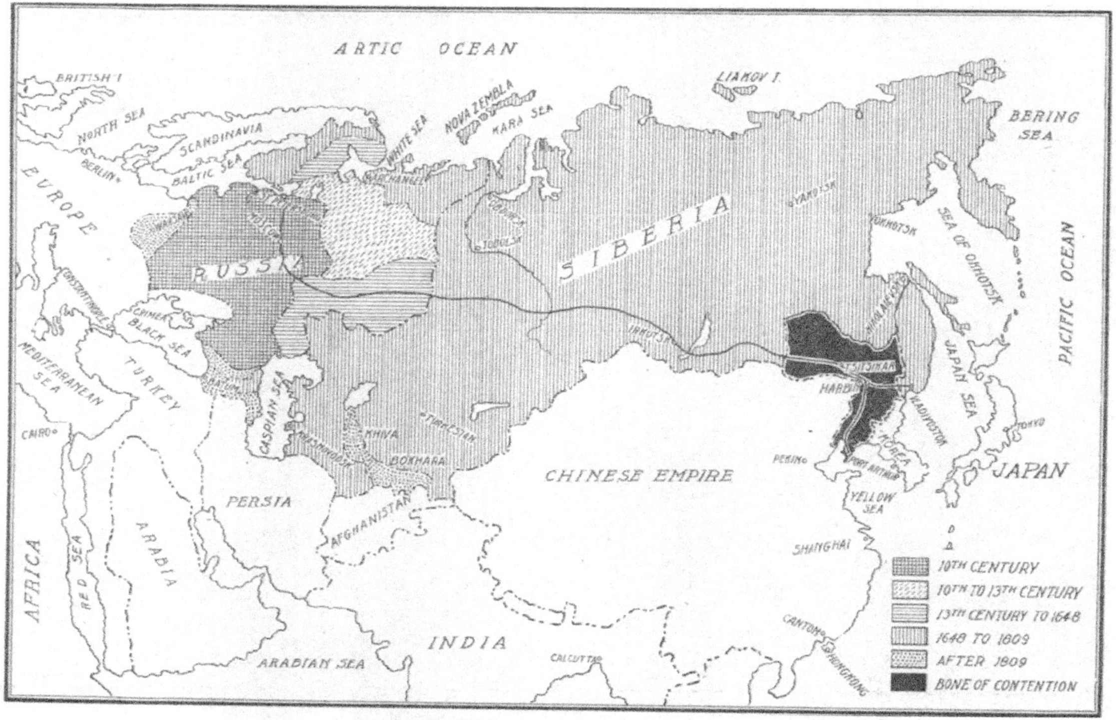

TERRITORIAL EXPANSION OF RUSSIA.

MAP SHOWING POINTS OF ACTIVE OPERATIONS IN THE WAR BETWEEN RUSSIA AND JAPAN.

Austria and the Balkan Peninsula, but only a small part of modern European Russia. The northeasternmost of these peoples, living in the district nearest the St. Petersburg and Moscow of today, were perhaps the nucleus of the present Russian Empire. All of these people, however, were divided into independent tribes, often warring among themselves, and constantly subject to attack from their enemies on every side.

Of all these invading enemies, the most warlike were the ones who came down from the northwest, the same who are known in western history as the Northmen or Norsemen. We do not need to be reminded what a large part they played in all western and northern Europe. In desperation the Slavs united—let us speak of them as Russians hereafter—and drove back the invaders into Scandinavia, whence they came. But as soon as the strangers were gone internal warfare broke out once more, and confusion was worse than ever. It was then that a strange event occurred. Two of the Russian tribes, recognizing the prowess of the Northmen, sent an embassy asking that princes be sent to govern them. "Our country is large and abundant, but there is no order," said the envoys.

It was in response to this invitation, in the year 862 A. D., that Rurik and two other Scandinavian chieftains came into Russia. Rurik reigned at Novgorod, to the south of St. Petersburg, and his companions established neighboring capitals. The latter soon died, and thus Rurik became the ruling prince of all the northern Russians. He left the throne to his descendants, and the great empire was founded. From Rurik the greatest and proudest of the Russian families of today claim their descent.

It must be admitted that there is a doubt in the mind of the historian as to whether this story of Rurik is a literal truth. Unquestionably he came as a ruler, but there are those who believe that he was also a conqueror; and that the tradition of the invitation grew up by the effort of the subjects to explain their subjection by an alien.

Six sovereigns succeeded Rurik and followed the pagan religion of their fathers, but Vladimir, the seventh in descent, who succeeded to the throne in 981, was converted to christianity. This faith was introduced by way of Constantinople by missionaries of the Greek church, who came northward in the year 955. Vladimir endeavored to make his own religion the religion of his people, and his success was so marked that before the end of his reign Russia was a Christian country, so far as its formal observances were concerned. Vladimir endeavored also to introduce Greek arts and sciences, and from Greece he procured architects and artisans to instruct his people in the various crafts. He was as great a warrior as he was a statesman, and his military conquests at one time embraced the whole of Poland.

Even this early in history, Russia was beginning to take a place among the nations, and three of Vladimir's granddaughters, the daughters of his son and successor, married the kings of France, Norway and Hungary. In those days it was the practice for sovereigns to divide their kingdoms among their sons. This practice was peculiarly favored in Russia, because of the semi-independence of the various grand duchies and important cities from each other. As time went on, however, the Russians discovered that they must unite for self-protection, and the strongest Grand Duke in each instance was able to dominate the situation by his own power. It would be of little interest to catalogue the successive sovereigns through the centuries of the Middle Ages. Almost the only historical facts that have been preserved about them refer to warfare, famines and great conflagrations that destroyed some of the cities.

It was in the year 1224 that the first Mongolian invasion threatened the Russians. The armies of Genghis Khan, who was a ruler of tremendous power and military ability, swept westward from his Central Asiatic dominions into Europe and carried everything before them. Even before that time he had taken possession of the west coast of the Caspian, and the lower course of the Dnieper river. The southern tribes of what is now European Russia, finding themselves overwhelmed by the Mongolian invaders, sent northward to their Russian neighbors and asked for help. The help was given, and a stand was made near the present town of Mariopol. The attacking armies were too strong, however, the Russians were completely routed, and nearly 90,000 of their army of 100,000 men were killed or captured. The Tatars swept northward unresisted, and put 50,000 of the inhabitants of Kiev to the sword. Fire and bloodshed marked their path, but finally for some unknown cause they turned backward and returned to their Central Asian deserts.

Thirteen years later, Bati Khan, the grandson of the first invader, came to the Volga with an army of 300,000 men, and once more desolated the succession of Russian cities which submitted to his arms. The Russian princes proved to be inadequate to stem the invasion, and failed utterly to raise and support armies for the defense of the country. Once again the Mongols turned back to Central Asia, even though their progress was meeting but slight resistance.

These successive invasions left the southern and central parts of the country in wretched condition. Taking advantage of the situation, a Swedish army came down from the north in 1242 to demand the submission of the country. They were met by a Russian army under command of Alexander, son of the reigning prince of Novgorod, and in a great battle on the banks of the Neva he won a victory which saved his country from a Swedish conquest and gained him the surname of Nevski.

One hundred years later Ivan the Second established the preeminence of Moscow as a city, and made it the capital of Russia, with himself as the ruler of all of the neighboring tribes and grand duchies. His successor, Dimitri IV, in 1380, met another invasion of Tatar hordes near the Don river, and defeated them with great loss, winning for himself the surname of Donski. Two years later, however, the Mongols again advanced, and this prince, betrayed by his allies, deserted the city, which was devastated by fire and sword until it was utterly destroyed. Before this disaster was repaired still another invasion threatened the capital, this time under the great ruler Tamerlane, or Timur the Tatar, whose name is among the most famous of all in Central Asian history.

These successive raids of Asiatic hordes, and others which have not been indicated, had left the country in a weakened condition, and the Tatars had taken possession of southern and eastern Russia without difficulty. They maintained capitals at Kazan and at Astrakhan, and until the end of the 15th century were a continual menace to the Russians of Moscow and the neighboring cities. Ivan III is known as one of the great rulers of Russia, because he reduced the Tatar city of Kazan, subdued the rival government of Novgorod the Great, an ancient republic, and finally destroyed and drove out the Golden Horde of Tatars, whose capital was at Astrakhan, at the mouth of the Volga. Until his reign all Russian sovereigns had paid homage and tribute annually to the Tatar conquerers, but Ivan put the ambassadors to death and fought the war that resulted to a glorious end and a Russian victory. This same powerful and ambitious prince of Moscow made treaties of alliance with and received ambassadors from the Pope, the Sultan, the Kings of Denmark and Poland, and the Republic of Venice. It was he who first assumed the title of Grand Duke or Prince of Novgorod, Moscow and All Russia, and added to his arms the double-headed black eagle, after his marriage with Sophia Paleologus, a princess of the imperial Byzantine blood from Constantinople. In fact, Ivan III may be called the true founder of the modern Russian Empire.

It was Ivan IV, the first monarch who took the title of Tsar, whose name has become infamous in history as Ivan the Terrible. He came to the throne in 1533, as a mere child, and not until twelve years later did his personality begin to make itself known. It was he who won the final victories over the Tatars at Kazan, and carried on various successful campaigns against neighboring monarchs to the north and west. An unsuccessful campaign against the King of Poland, and the death of his wife Anastasia, embittered him and the cruelties of his disposition became unrestrained.

As a crown to the many dreadful acts of barbarity of which he was guilty, he killed his eldest son with his own staff in a frenzy of rage, and died a prey to the grief and remorse which that crime occasioned, after having endeavored to atone for it by giving large sums of money to various monasteries.

Ivan the Terrible was a peculiar mixture of the liberal and the narrow, the cruel and the intelligent. He permitted Protestant churches to be built in Moscow for the foreign merchants who were trading there, but he never shook hands with an ambassador from his brother sovereigns of Europe without immediately washing his own hands when the visitor had taken leave. It was he who had the Gospels translated into the language of the people, and circulated freely throughout his dominions. As with many other monarchs his subjects prefer to remember his power and his conquests. Says a Russian writer, "The brilliant renown of Ivan survived the recollection of his bad qualities. The proofs of his atrocious actions were buried in the archives, while Kazan, Astrakhan and Siberia remained in the eyes of the nation, imperishable monuments of his glory."

One hundred years after the reign of Ivan the Terrible, Peter the Great came to the throne of Russia. The intervening century had been marked by civil wars with pretenders to the throne, by the murder of possible rivals by sovereigns or their next heirs, and by terrible famines in Moscow. It is believed that the famine of 1601 was the most appalling that ever devastated the capital of a country. Driven by the pangs of hunger, instances occurred of mothers having slain and eaten their own children. Men were entrapped into dwellings and killed and eaten. One hundred and twenty-seven thousand corpses remained for days unburied in the streets, and an eye-witness relates that 500,000 persons were carried off by the awful visitation. In order to alleviate these sufferings, the Tsar Boris broke open the granaries where food had been held at high prices by the avaricious monopolists of the day, and caused the grain to be sold at half its value.

It was the same ruler who established serfdom. In 1597 he

issued a decree forbidding peasants to leave the lands on which they were living. The application of this edict created serfdom as a fixed policy of the Empire, and it was nearly 300 years before that Russian form of white slavery was abolished.

The present imperial family of Russia, which we know as the Romanovs, was founded early in the 17th century, and strangely enough, as the result of an election. The internal dissensions which had weakened the country had made it the victim of foreign invaders. One Russian Tsar had been captured and taken to Warsaw, and the King of Poland had occupied Moscow. This aroused at last the latent patriotism of the people, and a stern campaign was carried on which resulted finally in the driving out of the Polish occupants. It was then, in 1613, that the military chiefs and landed proprietors met in assembly and elected as their Tsar Michael Romanov, son of the head of the church at Rostov, and then only 16 years of age. His rule was marked by an enlightened policy, and many liberties were secured to the people under the terms of the act by which he was given the crown. It was his grandson, who, in 1682, came to the throne, afterwards to be known as Peter the Great.

This greatest of Russian Tsars in the beginning was but half a sovereign, for he was crowned with his half-brother, who was to share the throne with him. Sophia, the sister of this partner of Peter's, was the Regent in actual control of the administration, but civil war soon rose between the various elements of this mixed government, and ultimately she was defeated and imprisoned, her brother resigned, and in 1689 Peter became sole Tsar, at the age of 17 years. The period of his reign, which ended with his death in 1725, is one of the most noteworthy in Russian history. The ruling passion of Peter was a desire to extend his empire and consolidate his power. The Russian advance toward the Pacific across Asia, and toward the Indian ocean by way of Central Asia, was planned by this great ruler, and even some of the details of the Russian threat against India are found in the records and archives left by Peter the Great. Much of his life was spent in war, and not always successful war, but his persistence was indomitable, and, by repeated return to the scene of his defeats, he usually succeeded in an ultimate victory. The Turks and the Swedes were his most inveterate enemies.

The works of Peter the Great in peace were as great as those in war. He founded his capital city of St. Petersburg in 1703, choosing a site for it which would enable him to look out on Europe, but through which Europe could not look into Russia. With all his energy, however, it is hard to approve of his judgment in placing his city where he did. It literally floats upon the islands of the Neva river, and is flooded every year when the spring thaws break up the ice. Peter possessed in an eminent degree a persevering mind and a resolute will which defied all difficulties. He formed and brought into a high state of discipline a large army; he established boat building and left his nation with a navy; he built canals and other works of public utility throughout his dominions, and he established commercial relations with China, and with almost every other country on the globe. He visited England and Holland but studied only their navies and expressed his abhorrence for the liberal principles of their government, which were, of course, diametrically opposite from his own absolutism. Great as was Peter, he was a man of violent temper and gross vices. He, too, killed his own son during a quarrel in the fortress at St. Petersburg, where the young man was imprisoned under suspicion of a plot to seize the crown and kill his father.

The name of Catherine is as great in Russian history as that of Vladimir, Peter, Nicholas or Alexander, and occurs almost as frequently in the person of various successive sovereigns. It was Catherine I who succeeded Peter the Great, and others of the same name have left their impress on history. It was not she, however, to whom the surname of the Great was given, but one who came seventy-five years later, at the end of the 18th century. The moral standards of the Russian imperial families had been retrograding, if such were possible, with the empresses rivaling the emperors for evil repute. Therefore it was no shock to the sensibilities of the nation when the greatest Catherine became also the greatest in corruption and licentiousness. It would be far from edifying to relate the details of Russian history during these centuries of intrigue, cruelty and imperial infelicity. We can only mention here a few of the essential historical facts which occurred during the reign of Catherine the Great. She extended the dominions of the realm southward and eastward, carried on war with the Turks, and formed a league with Sweden and Denmark. She also fostered the sciences, arts and literature, introduced important changes into the condition of the nobility and clergy, and began to organize a legislative commission, which would have been a great step toward liberty of thought. However, the embryo parliament early in its session commenced an inquiry into the evils of serfdom and the Empress promptly dissolved it.

Catherine was succeeded by her son Paul in 1796. He waged war against the French and the Italians, showed various eccentricities which made people doubt the soundness of his mind, and, finally, in 1801, his short reign was closed when a delegation of his generals strangled him to death in his own palace at St. Petersburg.

With the death of the Emperor Paul and the succession of Alexander, his eldest son, at the age of twenty-four, we reach what is the beginning of the modern era of Russian history. Alexander came to the throne in 1801, and in a very short time was involved in the Napoleonic wars which swept over all Europe. The Russians were defeated with their Austrian allies at Austerlitz in 1805, and after two years of negotiation and warfare Napoleon and Alexander met on a raft in the middle of the Nieman river and concluded an armistice which was a prelude to the treaty of Tilsit, making Russia the ally of France.

Two years later, in 1809, the alliance was broken and once more Alexander and Napoleon were enemies. Intermittent hostilities continued until the invasion of Russia by Napoleon in 1812. By many historians this is considered the beginning of Napoleon's downfall. With a splendid army he moved steadily northeast-

Courtesy of Everybody's Magazine.

RETURNING FROM THE REVIEW.

Czar and Czarina on Their Way to the Palace After Reviewing Russian Troops Ordered to the Front.

NICHOLAS II.
Czar of Russia.

CZARINA OF RUSSIA.

RUSSIAN EXILES TO SIBERIA.
Prisoners Sentenced to Lives of Horror in the Mines of Russia's Penal Settlement.

JAPANESE CAVALRY.
Unloading the Small Japanese Ponies Used by the Mounted Troops of the Mikado's Army.

ward, the Russians withdrawing before him, until finally on the 14th of September, the golden minarets and starry domes of Moscow came into the view of the French army. "All this is yours," said Napoleon to his enthusiastic followers. The shout of "Moscow, Moscow" was taken up by the foremost ranks and carried to the rear of the army. Hastening forward from Sparrow Hills Napoleon's army bivouacked in Moscow the same night, only to learn that the city had been evacuated not only by the Russian army, but by most of the inhabitants as well. Four days later the city was fired by the patriotic Russians, and in the terrible conflagration that followed the splendid capital was almost utterly destroyed. Violence and pillage added to the horror of the devastation, and when the flames died nothing was left of Moscow, says a Russian writer, save the remembrance of the city and the deep resolution to avenge its fate.

Napoleon found that an invading army could not live in Russia, nor could it meet and defeat the defending forces in a general engagement. Every effort made to negotiate with Alexander was rejected, and finally, on the 19th of October, the French warrior turned his back on Moscow with his army and his plunder, abandoning his empty conquest.

This retreat from Russia is a pitiful story of suffering. It is reckoned as one of the most complete disasters that ever befell an army. Cold and famine were enemies that could not be faced, and they were the allies of the Russian forces that hung on the flanks of the retreating army to harass them by day and night. The retreat became a rout and ended in utter confusion. The campaign against Russia began with an army of about 500,000 men. Of these less than 50,000 returned out of the general wreck safely to France. Of the others some 200,000 had been taken prisoners, 125,000 had been slain in battle, and as many more were dead from fatigue, hunger and cold.

So broken was the power of Napoleon by this catastrophe that the European combination against him at last became effective, and on the 31st of March. 1814, Alexander of Russia had the satisfaction of marching into Paris at the head of his army, which was one of the divisions of Allied Europe.

After the general peace of 1815 Alexander devoted himself to the internal improvement of his country, and his reign was marked by many judicious and liberal changes in the method of government and by a general advance in the welfare of his people.

During the reign of the Emperor Nicholas, which extended from 1825 to 1856, Russia continued to rise in prominence in international affairs. The period was marked by a gradual extension of the territories of Russia, southward and eastward into Asia, and warfare cursed several years of the reign. Hostilities were carried on successively with Persia, Turkey, Poland and Hungary. Most important of all, however, was the Crimean War, which began in the fall of 1853 and continued for nearly three years. Before it ended, England and France were drawn into the conflict as allies of Turkey, and the war is memorable for several names which have come into poetry as well as into history. The siege of Sevastopol, the Battle of Alma, the Battle of Inkermann, the storming of the Malakov and the Redan fortifications, and the Charge of the Light Brigade were events in this campaign never to be forgotten by students of military history. It was in this war that the older generals of the Russian army of to-day had their baptism of fire as young subalterns, and the lessons they learned then are being applied now to the war with Japan.

The Emperor Alexander II succeeded to the throne in 1856, and his reign of twenty-five years was likewise marked by warfare without, material improvement within his realm, and internal dissension among his people. Russians call him the Great Tsar Liberator, for it was he who, in 1861, declared the end of serfdom and the emancipation of more than fifty million white slaves. It is significant to note that this action was taken before the outbreak of our own Civil War, and was voluntary on the part of an autocrat. The emancipation was carried out peaceably, except for a few isolated outbreaks in remote regions of the Empire, where the terms of the proclamation were not understood.

The international history of Alexander's reign ended like that of his predecessor, with a significant war with Turkey. This latest of Russo-Turkish wars began in 1877 and continued for about one year. No other powers were drawn into hostilities, but in the settlement of terms of peace at the end of the war, Austria and Great Britain interfered in behalf of Turkey, to curtail the Russian claims, which would have made the Black Sea virtually a Russian lake. In this war, the names most famous among the list of battles are Shipka Pass, Plevna, Kars and Erzeroum. The men of greatest fame, whose names came into publicity at that time, were the Russian generals Gourko, Skobelev, Annenkov, Todleben and Melikov. Among the Turks, Osman Pasha gained a place in history for his remarkable military abilities.

It is in connection with the reign of Alexander II, the Tsar Liberator, that we may speak of the rise of the revolutionary sentiment in Russia. Such names as Nihilist, Anarchist, Terrorist, and the like, are generally arbitrary and do not always define what position is taken by those to whom the names are applied. The revolutionists of Russia did not begin by calling themselves Nihilists, but the name was applied to them from without. It is sufficient to say that a large class grew up in Russia, which desired the establishment of constitutional government and the end of the autocracy. The policy of suppression of free speech and free thought made their efforts to teach their sentiments objectionable, and even criminal. They were repressed with violence, and they used violence in return. Plots against the life of officials from the Emperor down, multiplied in number. At least half a dozen attempts were made to destroy the life of the Tsar. The Chief of the Secret Police of the Empire was assassinated on the streets of St. Petersburg. Young women students of the universities vied with young men in sharing the dangers of the conspiracy, and the actual assassination. Vera Zassulich attempted the life of the Chief of Police of St. Petersburg in his own house, and was acquitted of the crime by a jury. As a result a policy of the sternest sort was put in force toward the restless elements. At one time the railway train on which the Tsar was to travel was blown up by a mine placed under the track. At another time an explosion of dynamite under the dining room of the Winter Palace at St. Petersburg killed and wounded a large number of the guards who were there to protect the Emperor and his guests during a banquet about to be served. At last, on Sunday, March 13, 1881, as the Emperor was driving along the street, an explosive shell was thrown at his feet and he was carried home to die in an hour. It is necessary to study the further history of the Russian revolutionists in Siberia.

A natural result of the assassination of Alexander was the application of even stricter measures of repression against the revolutionists. The empire was declared under a state of siege, and martial law was put in effect everywhere. Universities were closed, civil courts found their jurisdiction almost gone, newspapers were suppressed, and the population of Siberia was augmented rapidly by the exile thither of large numbers of the most intellectual and advanced thinkers of European Russia.

It is declared by many writers who follow certain Russian historical authorities that it had been the intention of Alexander II to establish constitutional government and a parliament. They go so far as to declare that a draft of the constitution and a plan for a legislative body to be elected by the people were found on his desk, their promulgation interrupted only by his assassination. Justice demands, however, that doubt be cast on this assertion, however widely it may be accepted. The liberals of Russia utterly discredit the statement, claiming it to be but a Machiavellian method of arousing further sympathy for the amiable monarch himself, and further condemnation for the revolutionists.

Alexander III, who inherited the throne, and father of the present Tsar, had married a Danish princess, sister of the present Queen of England. The family connection between the reigning sovereigns of England, Germany and Russia, therefore, is very close, inasmuch as the sister of the present King of England was the mother of the Emperor William of Germany.

The reign of Alexander escaped the miseries of international war, and, indeed, was less marred by internal outbreak than some of its predecessors had been. The restlessness of the liberals was no less general, but it was subdued by the policy of repression in effect, and had less publicity therefore. This ruler died and was succeeded by his oldest son, Nicholas II, the present Emperor of Russia, on November 1, 1894. The young man came to the throne at the age of twenty-six years, and already has gained among his own people the title of Tsar Pacificator, because of his manifest personal desire for peace within his realm and without. It is a strange irony that in spite of this peaceful disposition, which is generally accredited to him as a man of gentleness rather than force, it should be his reign which encompasses the significant war between Occident and Orient in Eastern Asia.

After this rapid glance at the history of the Russian government and sovereigns, we need to observe the extension of Russian rule in Asia. The Asiatic territories of the Tsar are to be considered in three distinct divisions, and each has been gained by a method of its own and along a different line of advance. These may be characterized as the Russian territories in the Caucasus; the Russian governments of Turkestan, or Central Asia; and Siberia proper, with its later extensions south and east toward the Pacific. We of the United States have been far more interested in the latter of these than the others, but the diplomats of Europe have had to reckon with the Russian advance toward the Indian Ocean as not second in importance to the movement toward the Pacific.

The gradual conquest of the Caucasus resulted through successive hostilities with the rulers of Georgia, the ancient kingdom which included those regions, and with Turkey and Persia, the boundaries of which formerly included much that is now Russian territory. Roughly speaking, the Caucasus includes the region lying between the Black Sea and the Caspian Sea, traversed by a mountain barrier which connects these two bodies of water and forms the boundary between Europe and Asia. The ancient kingdom of Georgia itself was annexed to Russia more than a century ago, and the successive wars with Turkey have extended Russian boundaries southward until the entire east shore of the Black Sea is Russian, and the west shore of the Caspian is equally under the authority of St. Petersburg. An important railway extends from Batum, on the Black Sea, through Tiflis, the capital, to Baku, on the Caspian, the center of a tremendous petroleum industry. All this is Russian, and so dominant are the Russians along the southern shore of the Caspian and in Northern Persia that this great land-locked sea, five times as large as Lake Superior, is virtually a Russian lake. It is Germany and England that are most concerned about the Russian advance in this direction, for the Muscovites are aiming for a port on the Persian gulf, a railway from Baku through Persia to this port, and an entrance into the trade of the Indian Ocean and the East Indies.

The Russian conquest of Central Asia has been more spectacular and more rapid than was the conquest of the Caucasus. The region thus roughly characterized lies to the south of Western Siberia. It is bounded on the south by Persia and Afghanistan, on the east by the westernmost regions of the Chinese Empire, and on the west by the Caspian Sea. The conquest of Central Asia really began with the occupation of Tashkend, and since that event, in 1864, there has been hardly an interruption in the extension of the boundaries of Russia east and south.

No one knows when Russia and Central Asia first came into relations either peaceful or warlike, but there are records of Russian invasions of Khiva as far back as the beginning of the seventeenth century, in which the invaders were defeated with great regularity. A hundred years later the Khan of Khiva heard such wonderful tales of the power of Russia that he sent an envoy to Peter the Great with a request to be taken under his protection as a subject. Nothing came of that, but in 1717 Peter sent an expedition of 3,000 men to capture the tempting territories of which he had heard enticing tales. The expedition was destined to end in utter disaster, for successive detachments were waylaid by Khivans in ambush, and the Russians were exterminated to a man. Year after year punitory expeditions were sent from Russia, all ending with utter failure. There was some trade between Russia and the Khanates, mostly transacted at the great fairs of Russia, but whenever Russian travelers or traders went southward they were robbed, slaughtered or sold into slavery.

One of the famous expeditions against the Emir of Khiva was that under Gen. Perovski. He started from Orenburg, in June, 1839, with 6,000 men, 7,000 carts and 10,000 camels. One year later he reached Orenburg on his return march, with less than one-third of his original force, and with but 1,000 of the camels with which he had started. He had covered only half the distance to his destination, and had not even come in contact with the enemy he was to destroy. The deserts had defeated him.

The next advance of the Russians into Central Asia was made by a more eastern route, in the effort to reach Tashkend without passing through the Khanates of Khiva and Bokhara. This line of advance was more successful, and although progress was interrupted by the Crimean War, the northern invaders slowly but surely encroached on the territory of the rulers of the oases. Tashkend, the capital of the country, was taken in 1864, and four years later Samarkand, the Maracanda of Alexander the Great, fell into the hands of the Russians. The next territory to be absorbed was the rich valley of Ferghana, which was annexed to Russia in 1876, after a war with Khokand. This was the last extension of Russian authority southeastward into Central Asia until the absorption of the Pamirs in 1892.

The conquests which have been described did not give the Russians access to the Caspian Sea, for the region between the Khanates of Central Asia and that great body of water was held by a warlike race of nomadic tribesmen, allied with the desert which was their home, and very difficult to reach by invading armies. These were the Turkomans. Several campaigns in succession had failed to subdue them, when Skobelev was put in command of an expedition which was desired to be final. Russian histories are full of the accounts of glorious victories over the savages of the desert, but other histories of the same conquests call the same victories massacres, so that there is a considerable discrepancy as to the significance of the facts. At any rate the conquest ultimately reached the point where Skobelev was in command, and the Turkoman stronghold of Geok-Tepe was the only place of consequence which was holding out. This is but a few miles from the city of Askhabad, now the capital of the territory known as Trans-Caspia.

This fortress was the strongest fortification in Central Asia and was garrisoned by 35,000 Turkomans, who proved their ability as fighting men to the very end. The final assault on the ramparts was made on January 24, 1881. The Russians entered in force through a breach in the wall made by an explosion, and at last, after a desperate resistance, the defenders fled from their stronghold to make their way across the plain a few miles to the mountains on the Persian boundary, leaving 4,000 dead behind them. Skobelev ordered pursuit, with instructions to give no quarter. The infantry followed the fleeing multitude for seven miles and the cavalry for eleven. All who had not succeeded in escaping before that time, men, women and children, were killed in flight. In Skobelev's official report he stated that during the pursuit, after the assault, 8,000 of both sexes were killed, and he estimates the total number of Turkomans killed in the siege at 20,000. Lord Curzon, in his report of the affair, says that it was "not a rout but a massacre, not a defeat but an extirpation." This was the end of armed opposition to the assimilation of Turkomania by the Russian Empire.

Since that time, with few interruptions, the history of Russia in Central Asia has been peaceful. The railway which Generals Annenkov and Skobelev built eastward from the Caspian Sea to assist their military operations, has been extended 1,500 miles across the Bactrian desert and the rich oases, until it comes within actual view of the Pamirs, that tremendous mountain chain which marks the northern boundary of India. The easternmost terminus of the line is at Andijan, from which one may almost see India,

BAYONET EXERCISE WITH DUMMIES.
Russian Troops Practice a Charge Upon Oscillating Effigies of Foes.

A HOSPITAL CORPS.
Japanese Branch of the Red Cross Society.

KIAOCHAU, GERMAN NAVAL STATION.
Yellow Sea Port Where Wounded Were Taken Care of by Germans With the Permission of the Kaiser.

and the western territories of the Chinese Empire. A branch diverges from this main line and runs northward to Tashkend, the capital of all Central Asia. The line is operated as a military railway, and it is not yet freely accessible to travelers, but those who obtain permission from the Russian government in St. Petersburg have at their disposal one of the most interesting journeys in the world. In the course of the five-day railway ride from Krasnovodsk on the Caspian to Andijan and Tashkend in Central Asia, the traveler crosses the desert, sees ancient Merv, Bokhara, Samarkhan, Khokand, and half a dozen other cities of historic and romantic interest, crosses the two great rivers, Syr Daria and Amu Daria, the Jaxartes and Oxus of the ancients, and has opportunity to observe the remains of civilizations thousands of years old, that have changed but little since the dawn of recorded history.

Returning now to the northern line of advance by Russia, we find Siberia offering details of immense interest to American readers. There is a very definite parallel between Siberia as the Great East of Russia, and our own Great West of the United States. Each required courageous pioneers to advance by difficult marches, through trackless forests, across great rivers, and over mountain ranges, before cities could be built and commerce established in the remote regions. Siberia furnishes its picturesque conquerors to history, men who may be named in the list with Cortez and Pizarro, DeSoto and DeNarvaez. They found no high civilization awaiting them, with jewels and gold as a reward, no land of tropic beauty to delight them, but only a vast country sweeping north to the Polar Sea, inhabited by rude and primitive races, its climate harsh over large portions of the land, and its riches requiring labor to develop them. The gold of Siberia had not been washed from the rivers and formed into ornaments ready for the conqueror's hand, as was the gold of Mexico and of Peru. The furs were rich, but the getting of them meant long winters in the Far North. Agriculture was in the future, and was never considered as one of the inducements to conquest. In spite of what seem now to have been small temptations and great difficulties, the

Russian advance across Asia to the Pacific Ocean was the most rapid of all such progresses which history records, if its extent be calculated.

Yermak was the first Russian conqueror who left a great name behind him by his work to the east of the Urals, and to-day he is all but canonized by Russian historians. Others dispute as to his right to be called great, but there is no doubt that his work was in large measure the influence which added Siberia to the Russian possessions.

The first raids upon Siberia were made in the twelfth century by traders from Novgorod, who sought the valuable furs. But no settlement or permanent conquest was intended. Some hundreds of years later traders from Moscow made similar incursions across the Urals, and on the way they built huts, cultivated the land, and made small settlements on the European side of the mountains. In 1499, seven years after the first voyage of Columbus, the Muscovites sent an armed expedition, which conquered lands on the Obi River, and returned with many prisoners. This expedition brought back wonderful tales of the country and its people, which in course of time became interesting legends.

Next came the period of the Tatar hordes, who raided western Siberia, then called Yugra, just as they raided Russia itself. These early Siberians offered to pay tribute to the Russians on condition of being protected from the Tatars, and Ivan the Terrible, who about that time was victorious over the Asian invaders, accepted the offer with alacrity. About all he ever did was to collect the tribute, sadly neglecting the obligation which it incurred, of coming to the help of the hapless Siberians.

Cossacks under Yermak were the real pioneer conquerors of the land. Wild horsemen they were, and it was the pleasure of such lawless raiders to fight Tatars or anyone else who stood in the way. When at last Russian settlements were firmly fixed in the Urals, the family of Stroganov, now one of the greatest in Russian nobility, rose to be among the most rich and powerful of all the scattered settlers. By an imperial charter they were granted certain commercial and industrial monopolies, and exemption from taxation. In exchange for these privileges they were required to defend Russia from the incursions of the wild races beyond the mountains. For three generations the Stroganovs gained great wealth while thus defending their country. By this time they had learned the possibilities of the great region beyond, and they induced Ivan the Terrible, the reigning Tsar, to extend their charter and permit them to begin an invasion of Asia at their own expense.

Yermak had been a boatman on the Volga River, then a Cossack freebooter, and finally a river pirate, plundering vessels where before he had earned an honest living. At last the Tsar ordered that Yermak and his band be captured and hanged, and an army was sent to execute the order. The adventurers fled northeastward up the Kama River to the wild country in the edge of the Urals, where their leader had passed his boyhood. It was just then that the Stroganovs had received their charter authorizing them to attempt the conquest of Siberia. Yermak was the right man for the undertaking, and they, caring nothing for his past history, induced him to head the army of invasion. His lieutenants were the pirate leaders who had been his trusted men in the years past. His expedition was composed of 800 men, including a considerable element of the Don Cossacks, augmented by a rabble of other fugitives from justice, border ruffians of half a dozen races. These the Stroganovs armed and equipped, and the motley army set off for the conquest of Siberia.

It was on New Year's Day, 1580, that Yermak and his men started across the Ural Mountains into Siberia. They had the best equipment of the times, including light cannons, muskets and arquebuses. The invaders advanced almost without opposition, through the forests of the Urals and the Tobol River, but at last the Tatar rulers whose power was threatened, began to take alarm and attempted to make a stand. The whole country belonged to Kutchum Khan, an old and blind Tatar chief, the same who in his earlier days had put to death the envoys sent by Ivan to demand tribute. As was to be expected, Yermak was successful in all his battles. His muskets terrified all enemies, and none could withstand them. He proved himself untiring in energy and fertile in strategy, and his movement on Isker, or Sibir, the Tatar capital, on the banks of the Irtish River, was unchecked by a single disaster. The town of Sibir was taken on October 26, 1581, and the Tatar chief fled southward with the remnant of his forces. The Russians adopted the name of the town as the name of the surrounding country, and from that comes the name of the great land of Siberia.

After this great victory, Yermak sent his lieutenant back to Russia to offer to the Tsar the new land which he had conquered. Their Imperial master promptly pardoned the great freebooter all his former crimes, accepted the gift at his hands, and sent officers to assist him with a body of troops. Yermak's campaign continued for two years after this. He was uniformly successful in spite of treacheries, heavy losses, distressing winters, and sickness among his men. Blind Kutchum, the Tatar, never gave up the struggle. He renewed his fight with troops drawn from the armies of his southern allies in the Steppes. Yermak started with fifty Cossacks to meet the old warrior, but failing to find him, relaxed his vigilance one night and in the extreme of exhaustion pitched a camp on the banks of the Irtish River and failed to maintain a guard. This was on the night of August 4, 1584. In the middle of the night, during a blinding storm, Kutchum and his men attacked the sleeping camp. Every Cossack was butchered before he could rise, except one who escaped to tell the news, and Yermak himself. The conqueror fought for his life, but finding himself overwhelmed, dashed into the river in the hope of reaching one of the boats. The weight of his armor dragged him down to death, and there in the river his body was discovered a few days later, to be identified by its rich coat of mail, and the golden eagle on his breast.

Yermak must have been no ordinary man, though Russian historians may have glorified him too much and others may have gone to the other extreme in calling him nothing but a swashbuckling highwayman. At any rate, his name to-day is honored all over Siberia, in the highest and humblest homes, as that of a cherished national hero, and his exploits are the subject of numberless songs and legends.

The death of the first conqueror, who in his campaigns had covered the regions of the Tobol and the Irtish, with many smaller rivers, was a blow to the progress of conquest, but the government at Moscow could not afford to let the country rest as it was, and troops under new leaders were hurried in to take up the work where he had left it.

Gradually the line of block houses which served for forts was pushed eastward and southward, always following the rivers, which were the only avenues of communication. The Cossacks were the pioneers year after year, familiar as they were with a life of hardship, and with the methods of river travel as well as of fighting. They dragged their boats across the portages from the tributaries of the Obi to those of the Yenisei, and so reached the heart of the country by way of the intersecting river routes. Tobolsk was founded some fifteen miles from the destroyed capital Sibir, in 1587, and in the early part of the next century the Cossack settlements on the Yenisei were begun. Yeniseisk itself dates from about 1620.

Ten years later came the news of the discovery of another great river, the Lena, far to the eastward, on the banks of which lived another strange race, the Yakuts. The boldest pioneers hurried there, and in 1630, the catch of sables—for furs were the wealth that tempted—amounted to 2,000 skins. The town of Yakutsk, since then one of the more notable exile stations, was established two years later, and the new river with its numerous tributaries became a great highway of trade. None of these advances was made without war, but always the Cossacks conquered, and their progress was constant and irresistible. It was in the vicinity of Lake Baikal that the hardest fighting was met in the resistance offered by the Buriats, a Mongolian tribe, but that was overcome as the other opposition had been, and in 1651 the city of Irkutsk was founded.

In seventy years then the Cossack bands had penetrated from the Ural mountains eastward to the very center of Siberia, and northeastward even beyond Lake Baikal to the shores of the Lena River, and the country from the mountains thus far was added to the possessions of the Russian crown.

The Russian advance into Siberia, as far as just described, was made through lands belonging to semi-civilized races. It was not until Lake Baikal was passed in the eastward march that the Russian conquerors had to reckon with an organized government with a recognized civilization of its own. Not until Lake Baikal was passed did the Russians meet the Chinese, and the history of Russo-Chinese relations, instead of being entirely warlike, involves the struggles of diplomacy, as well as those of arms.

Fifteen years before the founding of Irkutsk rumors of rich valleys along the Amur River and its tributaries had enticed wandering Cossack adventurers to penetrate thither, and, indeed, one party had crossed Asia by a more northerly route and reached the Sea of Okhotsk. In 1643, therefore, the first Russian expedition to the Amur left Yakutsk, bound southward into the region bordering on the recognized territories of the Chinese Empire and itself inhabited by Mongolian tribes. This party, under the command of Poyarkov, descended the great river to its mouth, and returned a year later to receive high honor for the exploration.

Khabarov was the next commander to penetrate the region. He took with him strong bodies of troops, built forts along the river, fought battles impartially with native tribesmen or Manchurian soldiers from China, and gradually extended Russian power by strong measures. Russian villages began to grow along the Amur, a governor was sent to take charge of the remote settlements, and the Chinese took fright at the encroachment on their territory, for the lower Amur at that time was absolutely within the Chinese boundaries. In 1683 China began to send military forces to expel the intruders, and six years of intermittent warfare ensued, with varying results. At different times each power suffered severe losses, but in the end, by the treaty of Nerchinsk, signed August 29, 1689, peace was restored.

This first treaty of Russia with an Asiatic power is noteworthy also as marking almost the only instance in which Russians have withdrawn from territory once occupied. Under the terms of the treaty they were compelled to evacuate and destroy the forts and villages they had built on the Amur, and withdraw from the disputed region. It was more than 150 years before the Russians regained their foothold in the valley of the Amur, of which their dominance now is absolute. During this long interval Russian authority was extended unbrokenly from the region just east of Lake Baikal to the Sea of Okhotsk and Bering Sea, but this was a poor substitute for the milder and more fertile valley of the Amur, with its rich possibilities of trade with China. Repeated Russian embassies to Peking were welcomed or rebuffed according to the temper of the Chinese, but except for clandestine trade of small importance, no headway was made toward establishing friendly relations in Eastern Asia.

In 1847 the Tsar sent Count Muraviev to be governor of Eastern Siberia, and this energetic officer promptly showed a purpose to extend Russian power eastward along the Amur River to the Pacific Ocean. For six years successive exploring-parties studied the river, the coast line, and the islands north and south of the mouth of the Amur. Several Russian outposts were established on the lower course of the river and on neighboring islands. The outbreak of the Crimean war threatened the safety of these outposts, by the presence of French and English fleets in the North Pacific, and at the same time forbade the sending of supplies to the defenders by the long voyage around Africa and Asia. Here was Muraviev's opportunity. The relief of the settlements was a clear excuse, and the governor promptly organized a large expedition of soldiers, scientists and supplies, and with it floated down the Amur to the sea. Little Chinese opposition was shown, and from this time Russian occupation of the valley advanced rapidly.

By the time the Crimean war ended the British and French were besieging Peking to punish the Chinese for affronts received, and just as happened years later, all the powers took advantage of China's distress to enforce treaties at will. When the readjustment was concluded Russia found herself with treaties adding to the Tsar's possessions the north bank of the Amur River as far as the Ussuri, the whole of the Maritime Province of Manchuria between the Ussuri River and the sea, navigation and trade rights on the rivers of Manchuria tributary to the Amur, new consulates and trade privileges, and a rectified boundary extending Russian territory in Central Asia.

Slowly but surely the Russians have been extending their influence in Manchuria and in Peking ever since that time. Merchant colonies protected by Cossack soldiers have been established. Surveying parties and scientists have scrutinized the country in detail. The best books and maps of Northern China are those by Russian authorities. Finally when the utter defeat of China by Japan in the war of 1895 showed the weakness of Chinese defense, Russia found a new method of advance. The great northern power had found Vladivostok not quite satisfactory as a naval station, owing to the ice-bound condition of the harbor during several months each year, and had been casting longing eyes on the tempting harbors in the Korean coast line, although debarred from a seizure by Great Britain's attitude and strength. Victorious Japan obtained a treaty from China providing for the independence of Korea, the payment of a large indemnity, the cession of the rich island of Formosa, and finally the cession to Japan of the Liaotung Peninsula, on which is Port Arthur. Japan had captured this naval and military station by siege and assault, and held it at the time of the treaty-making.

At this point Russia stepped in, and secured the assent of France and Germany to a concerted protest against the Japanese occupation of any territory on the main land of Asia, on the plea that the integrity of the Chinese Empire must be preserved. It was futile to oppose the wish of these great powers, so Japan con-

Courtesy of Everybody's Magazine.

RUSSIAN IMPERIAL FAMILY.
Czar and Czarina With Their Four Daughters.

COUNT LEO TOLSTOI.
The Nobleman Who Espouses the Cause of the Russian Peasants, Works With Them in Barefeet and Denounces the Injustice of the Russian Government.

Courtesy of Everybody's Magazine.

RUSSIAN TORPEDO BOAT DESTROYER BUINI.

Courtesy of Everybody's Magazine.

RUSSIAN CRUISER SISOI VELTKY.

sented to forego the possession of Port Arthur, in return for an increased indemnity.

Next came the Cassini convention or treaty, negotiated at Peking by the present Russian Ambassador to the United States, then Minister to China, granting the right to build, maintain and protect a railway across Manchuria, connecting the Siberian Railway proper with the Russian naval station Vladivostok. Under the liberal interpretation of these clauses the Russians built cities and military posts all along the line of the railway, introducing thousands of soldiers into Manchuria, and making the northern part of the province, to all intents and purposes, as Russian as is Siberia.

It had not proved expedient, however, to show activities in Southern Manchuria, until in the fall of 1897 Germany seized the harbor of Kiaochau as a penalty for the murder of two German missionaries by the Chinese. This was Russia's opportunity. Immediately a Russian squadron moved into Port Arthur, and from that day to this it has been a Russian fortified naval station. Here begins the most bitter animosity of Japan to Russia. It was bad enough to be denied the fruits of victory in such fashion, but a hundred times worse to discover a little later that Russia in return for the intercession in China's behalf had herself been given that very Port Arthur for her own. Japanese will never feel that they have squared the matter till they have taken Port Arthur from the Russians, and made it their own. This will account in part for the pertinacity with which the Japanese continued their harassing campaign against that Russian stronghold.

The Boxer outbreak of 1900, centering around the besieged legations at Peking, gave to the Russians another excuse for enlarging their force and strengthening their position in Manchuria. At the same time it enabled the other European powers to obtain additional trade and railway-building concessions in the Chinese Empire.

After the first granting of Port Arthur to the Russians, the other powers demanded like favors from China, and England,

France and Italy in turn were given ports on the Chinese coast which they could use for military and naval stations in like fashion. This has helped to incense Japan, as another manifestation of the fact that Russia was but speaking for herself and not for China when declaring that the integrity of Chinese territory must be preserved.

Later Russian treaties with China have permitted the construction of the railway to Port Arthur, and by virtue of occupancy the Russians have been in a position to dictate trade conditions and almost all administrative details in Manchuria. Thus not only has Chinese authority in that province been virtually eliminated, but the interests of other powers, particularly the United States, England and Japan, have been seriously affected. For this reason American diplomacy has had a share in negotiations with Russia concerning Manchuria, although no clash has been imminent at any time. We are thus brought up to the period in the autumn of 1903, when a Russo-Japanese clash began to threaten as a result of conditions in Manchuria and along the coasts of Eastern Asia.

Let us now look at the Siberian Railway for a moment, not only as a factor of immense importance in the industrial and economic development of the country it traverses, but as a military and political factor in the present conflict. Russians long had recognized that a transcontinental railway was an imperative necessity, to bind their remote settlements on the Pacific to the seat of government and trade in Europe; to develop the agricultural and commercial possibilities of the tremendous areas of mid-Siberia; and to strengthen the military and political position of Russia in the Orient. In 1891 the construction of the line began at Vladivostok, with elaborate ceremonies and the turning of the first shovelful of earth by the present Tsar, then heir to the throne. Twelve years later, direct train service from Moscow to Vladivostok and Port Arthur was inaugurated, the line complete across two continents, for a journey of more than 6,000 miles, except for the short ferry across Lake Baikal. Within another year it is promised that this interruption will be ended by the completion of the line now building through the difficult country around the southern end of that great lake.

This is the most remarkable example of rapid railway construction known to the world. Except for the complacent treaty with China the result could not have been accomplished so readily, for the permission to construct the line across Manchuria reduced the total distance several hundred miles by cutting out the immense detour around the northern curve of the Amur River, and at the same time eliminated a vast amount of difficult construction along that river valley.

The Siberian Railway has been criticised for its light construction, but a careful examination of its track, roadbed, bridges, stations and equipment justifies the judgment that the line is practically as good as our own western prairie railways of twenty years ago. Low speed is maintained throughout the journey, however, and the equipment is not yet sufficient for the heavy demands upon it.

The cities along the line have grown rapidly since the railway was built, and have advanced in other ways by the increase of immigration and trade. Omsk, Tomsk, Krasnoiarsk, Irkutsk, Chita and Vladivostok in Siberia have felt the impetus; Harbin, Dalny and Port Arthur have become important Russian cities in Manchuria, and Blagoveshchensk, on the Amur River, off the line of railway, has grown rich and populous.

Along this railway the Russians have hurried their soldiers and munitions of war to the scene of hostilities. Its capacity may not be as ample as they would like to have it, but manifestly it has been a potent factor in strengthening their position in the Far East.

The railway has been an influence in gradually reducing the rigors of the exile system with which the name of Siberia is so intimately linked in every mind, and ultimately, no doubt, there will be an end put to the abominable practice. Publicity is powerful to destroy evil, and as foreign travelers and investors enter Siberia to develop the latent wealth in mines and commerce the exile system will become better known and more repugnant to the world, to its final downfall. Already milder measures are in effect, generally speaking, than those found by George Kennan, that faithful investigator who told the truth about Siberia to the world nearly twenty years ago. The railway makes the journey itself less a torture to the exiles than it was in the days of the terrible marches, and in many ways the severity of treatment has been modified.

The people of Russia are a sturdy, virile race, inured to hardship of climate and life, illiterate in the mass, devoted to the forms of their faith—the Greek Catholic branch of the Christian Church. They are frugal but improvident, slow to improve their methods of business, agriculture or industry, and loyal to the Tsar, the Church and the Nation. Of course these generalities apply but to the masses. Educated Russians dominate the national affairs, and maintain a society of manners not unlike that of other peoples. The advanced element of thinkers—the radicals—must be reckoned with, too, as a vital force in the Empire, comparatively few in numbers, but potent to do good when the leaven of liberty begins to make itself felt in the autocracy, and freer forms of government begin to be introduced. Whatever the result of the Russo-Japanese War, on international affairs and the two powers engaged, it promises to make for the ultimate benefit of Russia by forcing the Russians to look to themselves and their country for an uplifting of the people in education, liberty, and the good things in the civilization of other western nations, to which the Russian masses are so sadly deficient.

A few condensed facts in regard to the government, area, population, cities, religion, education, agriculture and commerce of Russia, will serve to close this rapid account of conditions in that great empire which sweeps across two continents, from Atlantic tidewaters to the Pacific.

The Emperor Nicholas II, Tsar of all the Russias, was born May 18, 1868, and ascended the throne at the death of his father, November 1, 1894. The government of Russia is an absolute hereditary monarchy. The whole legislative, executive and judicial power is united in the Emperor, whose will alone is law. There are, however, certain rules of government and methods of administration, which the sovereigns of the present reigning house have acknowledged as binding. Nicholas is the head of a tremendous system of autocracy, but it is greatly to be doubted if he could suddenly and radically alter the system, in spite of the power which pertains to his office. Many students believe that the Russian Emperor who should attempt to destroy the power of the autocracy, and free the nation from its present form of government, would promptly fall a victim to his own benevolent impulses.

The Russian empire comprises one-seventh of the land surface of the globe, with an area of about 8,650,000 square miles, or nearly three times the area of the United States without Alaska. No country in the world has gained more rapidly in population. In 1722 the inhabitants of Russia numbered 14,000,000. By the beginning of the 19th Century they approximated 40,000,000, and one hundred years later, at the beginning of the present century, the census showed a total of about 135,000,000 inhabitants. The population is by no means homogeneous, although its largest element is made up of Slavonic groups, known as Great Russians, Little Russians and White Russians. Within the empire, however, there are at least fifteen races represented with a total of more than 1,000,000 each, while Russian figures indicate nearly 150 races and tribes included within the immense sweep of territory.

The established religion of the Empire is the Greco-Russian, officially known as the Orthodox-Catholic faith. The Emperor is the head of the church, but he has never claimed the right of deciding theological and dogmatic questions. Practically the Procurator of the Holy Synod acts as head of the church administration. The points in which the Russian church differs from the Roman Catholic faith, are in denying the spiritual supremacy of the Pope, in not enforcing the celibacy of the clergy, and in authorizing all people to read and study the scriptures in their own language.

The school system of Russia theoretically includes a complete organization from primary grades to universities, for both sexes, with mining schools, agricultural schools, and other technical institutions. Practically, however, the organization is incomplete in the extreme and the facilities for education are not in reach of a large proportion of the children of school age. The restraints on the freedom of the press, too, and the illiteracy that is so widespread, hinder the publication of books and newspapers, so that the press is in a comparatively low state of development and progress.

Russia is primarily an agricultural country, with wheat, rye, barley and oats as the most important crops. In the southern provinces and in Central Asia cotton and tobacco are produced to great profit. Immense forest areas contribute great wealth to the industrial life, and the flocks and herds are another factor of great importance in the production of wealth. The soil of Russia is rich in ores of all kinds, and mining industry is steadily increasing. Gold, silver, lead, zinc, copper, iron, coal, naphtha and salt, are the most important products of the mine. Manufactories likewise are growing with great rapidity throughout the empire. Textile fabrics, paper and cardboard, chemicals, products of leather, china, glass, iron, steel and machinery are the most important manufactures.

The chief trade of the Empire is carried on through its European frontier and the Black Sea, and in all of the departments of commerce which would naturally develop out of the industries just mentioned, Russia is a rising factor in the trade of the world. The great river systems have been supplemented by canals in all directions, and by the rapid construction of railways throughout the Empire. If the undoubted energy of the rulers and the people, and the amazing resources of the realm, should be directed to the expansion of trade, education and modern development, instead of being wasted in the support of tremendous armies, Russia would rise even more rapidly into a position of pride among the nations of the world.

CHAPTER IV

JAPAN, THE ISLAND EMPIRE

BY TRUMBULL WHITE.

The Rise of an Asiatic Race into Full Fellowship in the Family of Nations—Historical and Descriptive Sketch of the Land and the People—The American Commodore Perry Opens the Country to the World—Growth of Western Civilization—The Mikado Restored to Power—End of the Feudal System—Constitutional Government Established—Japan's War with China and Its Far-Reaching Results—Extension of Commerce and Influence.

IN ALL the world there exists no parallel for Japan. Unique among nations is the Island Empire in her history, her manner of progress, and her relations with the other nations of the earth. Now that the eyes of all civilization are turned toward the Orient, with Japan as the focal point of observation, we shall find ourselves learning of a land of beauty, inhabited by a people of high intellect and noble spirit, with an inspiring past, a significant present and a future promising tremendous things.

In all the history of the world there has been no more picturesque event on the domestic side of international life than the entrance of Japan into fellowship among the nations of the western world. Within the span of half a century the Island Empire has passed from a reigning policy of stern isolation and refusal even to exchange civilities with other countries, to an intimacy of contact and a confidential friendship with the rest of the world hardly excelled between nations of common ancestry and life, and never before reached by those of such diametrically opposite birthright, traditions and customs.

In our sketch of the Russian Empire we have observed that power historically by the biographical method, studying successively the great figures whose personalities have stood out as factors in Russian affairs. We have seen how Yermak, Peter the

Great, Muraviev and Khabarov embody the significant facts of Russian history in Asia. But for Japan, it is well to depend on description rather than biography for the clearer view of affairs and conditions, past and present. Not that Japan has been lacking in national heroes and commanding personalities who stand out in the annals of the country. But so recent is our own interest in the details of oriental politics and progress that these names do not bear the personal equation to us, the graphic quality that is so essential in biographical history. Peter the Great, Ivan the Terrible, Catherine—these have no strange sound to our ears. But Jimmu Tenno, the Empress Jingo, Yoritomo, Hideyoshi—what signify these great names to an American reader? How many Americans do not know the name of Nicholas, Emperor of All the Russias, and how many know that of Mutsuhito, Emperor of Japan?

Furthermore, the significant period of Japanese history, so far as it bears upon our own affairs and the interest of the rest of the world, is circumscribed within a comparatively few years, while that of Russia runs for centuries. It is of Japan of to-day that we want to know, our demand is for a description of things as they are, the strange conditions and customs of life, the picturesque people and places, the introduction of western forms of civilization into a nation that was the embodiment of all that was oriental until the middle of the last century. Therefore, while we shall glance at the sequence of historical events in Japan with full recognition of their importance to the Japanese, we shall look at the Japanese themselves as the subject of greatest interest to us.

Marco Polo, the Venetian traveler, brought the first knowledge of Japan to Europe when he returned in the year 1295 from his wonderful travels in China. The Chinese had told him of "Chipangu, an island toward the east in the high seas, 1,500 miles from the continent; and a very great island it is. The people are white, civilized, and well favored. They are idolaters, and are dependent on nobody. And I can tell you the quantity of gold they have is endless; for they find it in their own islands." We

MUTSUHITO,
Emperor of Japan.

EMPRESS OF JAPAN.

CZARINA OF RUSSIA.
In Robes of State.

CZAR OF RUSSIA.
In Robes of State.

know now that there are some trifling details of Polo's description in error, but there is no difficulty in seeing that he had been well informed about the Island Empire.

The Chinese wrote the name of the country Chi-pen-kue, so Polo's Chipangu was pretty close to the fact. From the Chinese the Japanese took the name Nippon, and prefixing to it the word dai, meaning great, they made it Dai Nippon, the Japanese name of the empire. From this we have made the word Japan, or Japon, as the French have it.

Two races are included in the population of Japan, the Japanese proper, and the Ainos. It is believed that the latter were the aboriginal inhabitants of the archipelago, but only some 17,000 of them now remain, all of them dwelling in the north island of Yezo. These are the descendants of a warlike people who once lived farther south, and were driven north gradually, during prolonged warfare, by the race which now makes up the Japanese nation. Now the Ainos are a peaceful race, primitive, ignorant and unclean in their habits. They are sturdy people, however, and are unique among races in that they have an abundant growth of hair, which gives the men not only beards of unusual length and thickness, but a hairy covering over the body as well. Their religion, their arts and their industries are crude in the extreme, and they have no trade of any sort except barter with the Japanese. The Ainos are of no consequence in the life of Japan to-day, and are of interest only to the students of anthropology.

Let us then turn directly to the people who make up the mass of the empire, and who contribute that virility, alertness and receptiveness which mark Japan as among the most progressive of nations.

Mutsuhito, the Mikado, or Emperor, of Japan, traces his descent through one hundred and twenty-three generations to the founder of the imperial dynasty in 660 B. C. This first Mikado was Jimmu Tenno, and if Japanese records may be trusted, the dynasty which he founded is the oldest in the world. It is true that the ancient records contain much that is fabulous, mythical and exaggerated, but they may be accepted as a fair guide to the facts, and offer material of the greatest interest to the student. As in many other nations the early history confuses material and supernatural affairs to a considerable degree, and the immortals in the Japanese story of creation are the progenitors of the Emperors themselves. Even to-day this attitude is held by the people, although of course in weakened form, and the Mikados are regarded as semi-divinities. It has been interesting to note that the messages from the greatest Japanese commanders reporting victories over the Russian forces have credited all the success to the divine influence of the Mikado himself, rather than to their own sagacity and the bravery of their men.

In the Third Century of the Christian era the Empress Jingo took a Japanese army across the straits and conquered Korea. She is the greatest female character in Japanese history and tradition, equally renowned for her beauty, piety, intelligence, energy and martial valor. From her conquest of Korea came literature, religion and civilization to Japan, at least as far as Chinese models were to be adopted, and the Japanese of to-day are but following the example of their progenitors in seeking to dominate Korea, and in their willingness to accept and absorb whatever they find in other civilizations more serviceable than their ancient models. In this early day the Koreans were more learned than were the Japanese, and it was they who introduced the study of the Chinese language and the art of writing itself into the Island Empire. Artists, artisans and educated men were brought from Korea to instruct the people, new industrial methods were introduced, and successive emperors were zealous in encouraging the arts of peace. We may say, then, that this was the first of three great waves of foreign civilization which have entered Japan. The second was from Western Europe in the Fifteenth Century, and the third was from America and Europe in the decade following the advent of Commodore Perry, the latter a wave which has steadily gained in force instead of diminishing.

A peculiar governmental condition existed in Japan through

many centuries, ending only in recent years. There came a time when a weak Mikado was on the throne, with a strong and powerful prime minister, and out of this fact was perpetuated a sort of a double authority, in which the Mikado by virtue of his office and his descent from the gods, held the semblance of power, while the prime minister, or Shogun, as he came to be called, really had all the business of the country in his own hands. As time went on the Shoguns became more powerful, and the Mikados less powerful, until the latter, although held in the highest reverence, were little more than puppets. This system of government, called the Shogunate, ended only by a civil war, which broke out in 1864 and lasted for several years. During this time the country was torn by revolutions, but when the readjustment was at last complete, the Mikados were restored to actual power and the powerful families which had supported the Shoguns were compelled to accept the situation and become loyal subjects of their hereditary ruler.

During the early centuries of the Christian era, Japan and China kept up friendly intercourse, exchanging embassies on various missions. But in the Twelfth Century civil disorders in both countries interrupted the acquaintance, and when it was resumed friendship was not restored with it. By this time the Mongol Tatars had conquered the Chinese Empire, and the new Emperor, Kublai Khan, at whose court Marco Polo was then visiting, sent letters demanding tribute and homage from the Japanese. Repeated embassies making this demand were sent home rebuffed, until finally the Japanese in exasperation beheaded the last party of envoys. Then came preparations for war. An immense Chinese fleet sailed for Japan, but thanks to a tremendous storm that met them upon their arrival, the squadron was completely destroyed and the survivors were slain by the Japanese. In Japanese history this event holds much the same place as the destruction of the Spanish Armada in the history of England. This is the last time that China ever attempted to conquer Japan, whose people boast that their land has never been defiled by an invading army.

If the Japanese have not been invaded, they have not been delinquent in invading. The ambition of more than one Mikado has been to conquer Korea and even China. Korea more than once has been overrun by Japanese armies, even partly governed by Japanese officials, and on different occasions has had to pay tribute to Japan in token of submission. Japanese pirates were for six hundred years as much the terror of the Chinese and Korean coasts as were the Danes and Norsemen of the shores of the North Sea. A strong party in Japan has long held that Korea is properly a part of the Japanese Empire, by virtue of the conquest made by the Empress Jingo in the Third Century, and by the Mikado Hideyoshi, in the Sixteenth Century.

During these centuries leading up to the first contact of the Japanese with European civilization, the Island Empire was rising rapidly in strength and prosperity, and the people were displaying the same qualities which make them noteworthy now. Arts and sciences flourished and developed. The spirit of military enterprise and internal improvement was alive. Contact with foreigners of many nations awoke a spirit of inquiry and intellectual activity, and on the seas the Japanese proved even in that day that they were capable sailors and energetic adventurers. The Japanese ships built in the Seventeenth Century were larger and better than the Chinese junks of to-day, superior in size to the vessels of Columbus, and nearly equal to the Dutch and Portuguese galleons of the same time. They were provided with artillery, and a model of a Japanese breech-loading cannon of that date is still preserved in Kioto. Voyages of trade, discovery or piracy were made to India, Siam, Burma, the Philippine Islands, Southern China, the Malay Archipelago and the Kuriles.

The Japanese were not always as hospitable to foreigners as they are to-day. The Shoguns, who were in actual authority, never permitted foreigners to negotiate directly with the Mikado, but dealt with embassies themselves at their own will. It is believed that the first European who landed on Japanese soil was a Portuguese adventurer named Mendez Pinto. He came with a pirate trader in 1542, and returned to China loaded with presents. The new market attracted hundreds of Portuguese adventurers to Japan, who found a ready welcome. Missionaries followed merchants from India, where the Portuguese already had a prosperous settlement. For a time the Catholic missionaries were given every facility, and in forty years there were two hundred churches and 150,000 native Christians. Before long the different missionary orders, Jesuit, Franciscan and Augustinian, began to clash, as a result of the political and religious wars then almost universal in Europe. All the foreigners were slave-traders, and thousands of Japanese were bought and sold and shipped to China and the Philippines. The seaports were the resorts of the lowest class of adventurers of all European nations, and the result was a continuous series of uproars, broils and murders among the foreigners.

Such a picture of foreign influence and of Christianity, as the Japanese saw it, was not calculated to make a favorable impression on the Japanese mind. Finally an edict was issued commanding the missionaries to assemble for expulsion from Japan, and a concerted effort to crush out Christianity was made. Churches were burned, and missionaries and their converts were slain. An edict forbade the exercise of the Christian religion, and persecution was carried on in every violent form. The exiled missionaries kept secretly returning and sentence of death was pronounced against any foreign priest found in the country. Fire and sword were used to extirpate Christianity, and to paganize the same people who had been converted in their youth by means hardly less violent. Thousands of the native converts fled to China, Formosa and the Philippines. The Christians suffered all sorts of persecutions and tortures that ingenuity can devise. If anyone doubts the sincerity and fervor of the Christian converts of to-day, or the ability of the Japanese to accept a higher form of faith, or their willingness to suffer for what they believe, he has but to read the accounts of the various witnesses to the fortitude of the Japanese Christians of the Seventeenth Century.

When this persecution ended, foreign trade had been annihilated, contact with the rest of the world had ended, and Christianity in the Island Empire had been virtually blotted out. It was a peculiar retrogression that has not been equaled in any other country, so far as history records.

The English, like the Portuguese, attempted to open trade with the Japanese, but with no success. Will Adams, an English pilot, and the first of his nation in Japan, arrived in 1607 and lived in Tokio, then called Yeddo, until his death thirteen years later. He was treated with kindness and honor, and he became a very useful man, owing to his knowledge of shipbuilding and foreign affairs. There are still living Japanese who claim descent from him, one of the streets of the city was named for him, and the people of that street still hold an annual celebration in his honor. During these centuries of isolation, there was no contact with the rest of the world except by way of the Dutch trading vessel that was permitted to call annually at the little Island of Deshima in the harbor of Nagasaki. Here a few Dutchmen were permitted to live under degrading conditions of restraint.

It was the American Union which opened the door of Japan to western civilization. The United States, in common with European powers, very much desired access to Japanese ports. Supplies were frequently needed, particularly water and coal, but no distress was ever considered a sufficient excuse for the Japanese to permit the landing of a foreign vessel's crew. Shipwrecked sailors frequently suffered great trial and danger before they were rescued and restored to their own people. Even Japanese sailors who were shipwrecked on other shores, or carried out to sea, were refused readmission to their own country when rescued by foreigners. Commodore Matthew C. Perry, of the American Navy, and President Millard Fillmore, consulted together and with the advisers of the President, and decided to enforce an entrance and a treaty upon the exclusive empire. A fleet under the command of Perry was assigned to the undertaking, and on the 7th of July, 1853, four American warships appeared in the Bay of Yeddo. The local officials promptly notified the "barbarian" envoy that he must go to Nagasaki, where all business with foreigners had to be done. The barbarian refused to go. He informed the messengers that he was the bearer of a letter from the President of the United States to the Emperor of Japan, and that he was going to deliver the letter. Alarm spread rapidly along the Japanese shores, but at last, after eight days, Perry's patient insistence and the demonstrations made by the fleet, impressed the Japanese, who had never seen a steamboat before, and won success for Commodore Perry's mission. A Japanese commissioner came to the landing, a magnificent pavilion was prepared for the ceremonies, and with great pomp and ceremony the Americans landed and delivered the letters and presents from President to Mikado.

Six months later Perry returned with a much larger fleet, and on the 31st of March, 1854, a treaty with the United States was signed. A few months later treaties were entered into with all the leading powers of Europe, and it is universally recognized by the Japanese, as well as by Europeans, that the credit for opening the Japanese door to western civilization belongs to the United States and its diplomat-sailor, Commodore Perry.

Even after treaties of friendship were signed and legations were established in Japan, bitter enmity existed against foreigners. American and European officials attached to the embassies were murdered in broad daylight on the open road, and it was necessary to enforce penalties and punishments upon Japan for these crimes. Japanese ports were shelled and indemnities were collected on more than one occasion, by American, British, French and Dutch fleets. However, as time advanced and acquaintance grew, the Japanese came to understand that they must maintain peaceful and friendly relations with the nations of Europe and America. Japanese travelers visited the rest of the world, and saw what western civilization did for its people. The Shogunate was abolished, as we have already seen, and under the direct authority of the Mikado conditions improved rapidly. The feudal system, likewise, which had followed forms very similar to those known in Europe during the Middle Ages, survived in Japan until 1871, and with its termination came another noteworthy advance in national affairs.

In 1872 the Japanese challenged the admiration of Christendom by making a stern fight against the coolie traffic, which was really nothing more than the slave trade. This had been carried on by the Portuguese at their little colony of Macao on the Chinese coast, from which they decoyed thousands of Chinese yearly to be shipped to plantations in Cuba and South America. They had extended their nefarious operations to the coast of Japan, and finally by the energy and persistence of the Japanese the trade was abolished by this pagan nation, although the foreign consuls and ministers in China and Japan, with the exception of those of Great Britain and the United States, protested against the Japanese action and defended the slave trade on the grounds of necessity.

In the last thirty years the record of Japan has been one of constant and steady advance into fellowship with other nations in its foreign affairs, and noteworthy improvement intellectually and industrially among its own people. Railways, telegraphs, a lighthouse service and a navy have been constructed. Two national expositions have been held with great success. A war with China has been fought to a triumphant conclusion, affording proof of the strength of Japan and the weakness of China in military and naval affairs. The details of this war will be related in the ensuing chapter on Korea, which was the battlefield between the two rival nations. The liberty of the press and liberty of speech have been established. A parliamentary government has been created voluntarily by the Mikado, who thereby relinquished his inherited despotic rights, and made Japan a constitutional monarchy.

Perhaps the most noteworthy evidence of the entirety with which the nations of Europe and America have accepted Japan as a power among powers is the abolition of what is known as "extra-territoriality" in 1899. Until that time, foreigners residing in Japan were not subject to the jurisdiction of the Japanese courts. Lawsuits between foreigners, or in which a foreigner was even one party to the suit, were tried before courts maintained in the con-

OFF TO THE WAR.
A Japanese Soldier Bidding Farewell to His Family.

Courtesy of Everybody's Magazine.
MARQUIS ITO.
Japan's Greatest Statesman.

Courtesy of Everybody's Magazine.
FIELD MARSHAL YAMAGATA.
The "Right Hand of the Mikado."

Keeper of Dogs on Way to the Front.

USE OF DOGS IN WAR.

Dog Guiding Ambulance Corps to Wounded Soldier.

GENERAL VIEW OF VLADIVOSTOK.
Russia's Chief Seaport and Naval Station on the Pacific.

VIEW OF CHEMULPO.
Principal Open Port of Korea Where Two Russian Cruisers Were Sunk by Japanese Warships.

sulates by the different nations having such offices in Japan. The same condition existed in the event of criminal offenses, and all this was true because Europeans and Americans did not trust the Japanese with the administration of justice to foreigners who might be residing in the islands. This, of course, was exceedingly distasteful to the Japanese, and for years they had persisted in the effort to eliminate the objectionable clause from the treaties. At last, in 1899, a new series of treaties was made by which Japan entered into the exercise of full authority over all people within her boundaries, in just the same fashion as such authority is enjoyed by the United States or by England. It is interesting to note that as a matter of fact, in the litigation that has risen since that time, and in the trial of foreigners for criminal offenses against other foreigners, or against Japanese, the suits have been conducted in an orderly fashion, fully preserving the rights of all parties, and promoting justice in a judicial manner, just as truly as before. This is the universal testimony of foreigners living in Japan.

Nearly four thousand islands make up the Empire of Japan, but only four or five of them are large enough to give them much importance, and around these a cordon of defense is formed by the reefs and shallows and intricate channels of thousands of islets. Until recent times the Island of Sakhalin was included in the Empire, but a treaty with Russia joined that island to the possessions of the Tsar, since which time it has been a settlement for criminal and political exiles from Russia. In exchange for this the Russians yielded the Kuriles, a group of islands extending north toward Behring Sea, to Japanese authority, and these now form the most northern extension of the Empire. The exchange was never quite satisfactory to the Japanese, and they have always intended to resume possession of Sakhalin, if they are ever able to do so. The southernmost possession of Japan is Formosa, which was taken from China as a part of the spoils of victory after the end of the Chino-Japanese war of 1894. This is a large and rich island, the resources of which are as yet but little developed, and the inhabitants, except on the coasts, still in a state of savagery.

The area of the Japanese islands, including Formosa, is nearly as great as the New England and Middle Atlantic Coast states, but of this, more than one-half consists of mountain land, much of it still lying waste and uncultivated, although apparently capable of tillage. The shores rise abruptly from the sea, and there is a gradual rise until the mountain backbone, which extends throughout the island chain, is reached. The highest peak is Fuji-yama, which rises to a height of more than 12,000 feet above the sea. It is a beautiful mountain, and is the first glimpse one has of land when approaching Yokohama from the United States. Japan's northern limits correspond approximately with Paris and Newfoundland, while the southern ones are on the latitude of the Bermuda Islands, and of Cairo in Egypt. Coming nearer home, it corresponds pretty nearly in latitude with the eastern coast line of the United States, added to Nova Scotia and Newfoundland, and the contrasts of climate between Newfoundland and Florida are no more remarkable than those observed between the extreme northern and southern regions of Japan. Even this wide range of latitude does not include the outlying island of Formosa, which is genuinely tropical, corresponding in latitude to Cuba and the Hawaiian Islands.

Thanks to the ocean currents, which move northward from the Malay archipelago and the South Pacific, the climate of Japan is moderated to a delightful degree, and the rains which are so essential to vegetation are assured at all seasons of the year. As a result of these facts and the fertility of the soil, Japan is one of the most productive of lands. It must not be thought, however, that the country is tropical. North of Tokio frost and even heavy snowfalls occur in the winter. Earthquake shocks throughout the island are frequent, but of late years there have been none of great severity.

The bamboo flourishes in all parts of the island, sugarcane and cotton grow in the southern part, and tea is produced almost everywhere. Tobacco, hemp, corn, rice, wheat, barley, millet, buckwheat and potatoes are cultivated for the market. Mulberry for silkworm food is one of the important products. Forest trees and flowers of great beauty and variety are found everywhere, and many of them have been transferred to our own hothouses and gardens. The azalea, camelia, wisteria, cryptomeria, calceolaria and chrysanthemum are indigenous to Japan. In ancient times two species of dwarf elephants existed in the plains around Tokio, although they are now extinct. Monkeys, foxes, wolves, bears, antelopes, and various species of the deer family are found wild, while the sea is especially rich in seals, sea-otters and whales.

Thanks to the genial climate, the fertile soil, and the characteristics of the people, Japan has always been a country of industry, prosperity and thrift. The cities are large, and architecturally interesting, although, of course, entirely unlike anything known to the western world. Tokio, the capital, has a population of more than a million, while the other important cities, Nagasaki, Yokohama, Hakodate, Hiogo, Osaka and Hiroshima are important centers of trade and industry. All of these have foreign settlements, where American and European merchants have stores, banks and steamship offices, with churches and newspapers and society of their own. Yokohama is the most important of these foreign settlements, and it is the great mercantile center of American and European trade in Japan. Commerce between Japan and the western nations increases year by year, with England standing first in the volume of trade, and the United States second and rapidly gaining.

The Japanese are not only industrious in their agricultural and commercial enterprises, in their mining and railway building, in their shipbuilding and their government, but they are at the same time equally faithful to the demands of intellectual life and the enjoyment of art and literature. The bronzes, the lacquer work, the ceramic arts, the silks and the paintings of the Japanese are recognized throughout the world as embodying some of the highest principles of genuine art, and every traveler is impressed with the beauty of the local products in these directions. Whatever they do, they do with the same attention to detail in the effort to reach perfection that they show in the planning of a military campaign or the administering of their governmental affairs. Their literature and their music have not taken forms that appeal to Americans, because of different standards of beauty and taste that rule, but the merit of what they do in these directions must be recognized by any careful student.

The people themselves are in a state of transition, as truly as is the government, and there are those critics who declare that their quickness of imitation is an evidence of weakness and dependence, rather than of originality. However, this seems to be an unfair judgment. It is true that they show an extraordinary capacity for change, and that they are versatile in the extreme, but they select with care to take what is best in the examples of other nations before them.

The average Japanese is frank, honest, faithful, kind, gentle, courteous, confiding, affectionate, filial and loyal; but love of truth, chastity and temperance for their own sake are not characteristic virtues. The male Japanese is more chivalrous to women than any other Asiatic. In reverence to elders and in obedience to parents the Japanese set a high standard, and indeed have developed filial obedience into fanaticism.

Physically the people are not unlike the Spaniards and the inhabitants of the south of France. They are of low stature, but well proportioned and strong, with an immense capacity for endurance of labor on small supply of food, shelter and comfort.

Japan is the paradise of children. There is no country where children are more devotedly loved and considerately treated than in the Island Empire. Playthings are everywhere, and holidays for the distinct benefit of the little folks are very numerous, upon which occasions the whole nation turns out to help the children have a good time. Even the adults have for their sports many of the plays and games which are left to children with us, and kite-flying is almost a national habit.

The dwelling houses are well adapted to their manner of life, except that they are not always built for sufficient protection against severe cold. The houses contain but little furniture, for the people do not sit on chairs, nor do they have high beds or tables. Sliding partitions and screens are everywhere, so that rooms may be divided and sub-divided almost at will. In the walls are recesses with sliding doors, into which the bedding is thrust in the daytime, to be spread out on the matting floor at night when it is time to retire. These mats answer the purpose of all ordinary furniture, and take the place of our chairs, tables and beds. They are made of rushes, or rice straw, to a thickness of three inches, and are soft to the touch. People never soil them with their shoes, but always walk barefooted about the house. In all their manners of life they are the most orderly and cleanly of people, and bathing is the most constant and the most universal to be found in any country in the world. Every house has its garden, and the garden is used as a place for dining as well as for recreation in pleasant weather. Outdoor life is greatly favored, and the houses themselves are built so openly that the foreigner feels himself almost out of doors even when he is under shelter.

The religion of the country includes two different systems, one known as Shintoism, which is the ancient religion of the people, and the other Buddhism, which swept over Japan after it was introduced from southern Asia. As an evidence of the receptivity of the people, however, Christianity has made remarkable progress, and religious toleration makes the work of the missionaries at the same time effective and congenial.

The women of Japan are rising as steadily as are the men. They are allowed a degree of freedom that is noteworthy in Asia, where generally women have little respect and consideration. Public and private schools are everywhere, and are well attended. No women excel the Japanese in the innate love of beauty, order, neatness, household adornment and household management, while in maternal affection, tenderness and faithfulness, Japanese mothers need fear no comparison with those of other lands. They direct the education of their children and have a degree of authority in their own households that places them in a position of great influence and dignity. The Japanese maiden is bright, intelligent, interesting, modest, lady-like and self-reliant. What the American girl is in Europe, the Japanese maiden is among Asiatics, and it is the appreciation of this fact, and of a certain sympathy between the people and the countries, that makes Americans so favored in Japan, and makes every American so pleased with what he sees in the Japanese Empire.

The Japanese have welcomed the phrase "the Yankees of the Orient," which has been applied to them by many travelers, understanding as they do that a distinct compliment is intended by the characterization.

Half a dozen lines of steamships ply between the American coast and the Japanese ports, and at least two of these lines are owned and operated by the Japanese themselves. The passage across the Pacific requires from fifteen to twenty days, according to the vessel and the route selected. The journey is becoming a more popular one every year, and travelers to the Island Empire always find awaiting them a welcome from the people, and manifold pleasures in the beautiful country itself. Now that the Hawaiian Islands and the Philippines are under American government, the distance between Japan and our outlying ports is greatly reduced, and we find ourselves dealing with Oriental peoples who in some measure are kindred to the Japanese themselves. A closer contact, a better understanding and more intimate personal and trade relations are certain to redound to the benefit of both countries, and we should cherish the undoubted friendship that exists between these two powers that look toward each other across the Pacific.

CHAPTER V

KOREA AND MANCHURIA

BY TRUMBULL WHITE.

"The Land of Morning Calm"—Sketch of the Hermit Nation—History, Geography, Resources, People, Customs—Weakness of the Government—The Scene of Jealousy and Strife—Rivalry of Chinese, Japanese and Russians—Manchuria and Its Relation to the Chinese Empire—Chinese Ports in the Grab-bag for Ambitious Nations—Secret Treaties with Russia—The Manchurian Railway—Port Arthur in Russian Control—The Boxer Uprising—The Looming of the War Cloud.

WITH a record of retrogression uninterrupted for centuries, at frequent intervals the object of jealous rivalry, and then the battlefield of warring neighbors; today, as in the past, the prize over which diplomates wrangle and armies clash, that poetical phrase, "The land of morning calm," applied by Koreans to their unhappy country, becomes an irony indeed. For truly, the little kingdom—or empire, as its rulers call it, with proper imperial pride—has fallen to low estate among the nations of the earth.

The miseries of the country, and its inability to raise itself out of the slough of despond into which late centuries have brought it, are the more conspicuous because of its proximity to Japan, that land of a kindred people, with resources and natural conditions not altogether dissimilar, where progress has been as noteworthy as its converse has been in Korea.

And yet it is a historical truth that much of the best in Japanese art, letters and industry came to the Island Empire from the Koreans, either being original with the people of the peninsula, or first absorbed by them from the Chinese with whom they were in intimate contact. Korean artists taught the Japanese to make some of the most treasured specimens that our travelers bring home with them from their journeys in the Far East. Korean scholars helped to create the language and literature of Japan. Korean artisans lent their skill to found Japanese industries that remain today almost in their original form.

Not that these benefactions were always voluntary. In large degree they came as a result of Japanese invasions of Korea, whence the returning armies brought what they saw that might be of service. But in the past, as in the present, the Japanese were alert, apt, intelligent, able to imitate, adopt, improve and apply their borrowed knowledge, be it handicraft or otherwise, so that the western nations of today are but repeating the experience of Korea in giving of their best to their island neighbors. These are the facts frequently cited by critics of Japan who declare that she has nothing good but what is borrowed, and that this very quality of prompt adaptability which we call capacity for progress, is an evidence of racial weakness, instability and unresourcefulness. Such a broad question cannot be discussed or settled here.

While Japan was gaining ground in every way, materially and intellectually, Korea was losing. It would be impossible to say that Korea lost because Japan gained, for the imparting of knowledge to another never harms but rather helps the one who gives. Nevertheless one fell while the other rose, and it is long since Korea could assume any place of credit in the family of nations. In some details there is a similarity between the retrogression of Korea and what we are taught to call the decay of China. But the parellel is not complete. The Chinese Empire by virtue of her immense area, her tremendous population, and her great trade commands certain forms of respectful consideration in international affairs even from those who are most intolerant of her failings. Korea being smaller, less populous, and of minor importance in the world's commerce, loses even this element of respect, and only as the peninsula becomes a factor in the affairs of other countries as a bone of contention, a part of the Far Eastern question, do we regard it seriously.

Korea in its extreme dimensions measures nearly the same length from north to south as does the State of California, and lies

THE LANDING PLACE, TOBOLSK, RUSSIA.
Here the Exiles for Siberia Get Their Last View of Freedom on the Way to the Mines.

SCENE IN A RUSSIAN PRISON.
Convicts With Half-Shaven Heads to Identify Them if They Escape.

but a little north of the same parallels of latitude. Its area, however, is but little more than half as much, or about 82,000 square miles. If a comparison on the Atlantic coast be preferred, Korea would extend from Portland, Maine, to Wilmington, North Carolina, with an area greater than the total of New Hampshire, Vermont, Massachusetts, Rhode Island, Connecticut and New York. Its population statistics are unreliable in the extreme, but the best estimates make the number of inhabitants about 11,000,000, or not far from the combined population of all the New England states and New York.

Measured by our standards of density this would seem to be a fairly populous country, but so far short of its possibilities does it fall, that from the oriental point of view it is sparse indeed. Japan, China, and the Indo-Chinese countries far exceed this in their density of population. Furthermore, it is clear that if the present primitive methods of agriculture and industry can support this number in Korea, the introduction of modern methods of intensive cultivation, and modern transportation systems, would immensely increase its capacity. Here we have one stimulus for the interest Japan displays in Korean affairs.

On three sides, west, south and east, Korea is circumscribed by the sea, joining the mainland of the Asiatic continent on the north, or rather on the northwest, with the Manchurian province of the Chinese Empire as the neighbor across the boundary line. For a few miles at the extreme northeastern corner of the country, also, Korea comes in contact with the Russian Empire, there being a common boundary line for a short distance from the sea. On the southeast, the islands of Japan come within a few miles of the Korean coast, so that the three empires of China, Russia and Japan are the contiguous powers. Such political and racial differences naturally breed jealousies, controversies and ultimate disorder.

Until recent years the Koreans have maintained a policy of exclusion against the rest of the world, so that little has been known about the strange people who inhabit the peninsula, except through the scanty information coming to us from Chinese and Japanese sources. Indeed, it was not known even by name in Europe until the Sixteenth Century, although Arabian merchants trading to Chinese ports had crossed the Yellow Sea and visited the peninsula. Korean youths also had been sent to study at Nanking, the ancient capital of China, where they may have met the merchants of Bagdad and Damascus.

The Koreans are in ignorance of the origin of their race, but it is conjectured that they come of a fusion of tribes, partly Mongolian, from the North, and partly Caucasian, from Western Asia. We class them now as kindred to the Chinese and Japanese, but they are distinctly different in physical type and in temperament.

For our purposes it would be fruitless to relate the centuries of strife during which the Koreans endeavored to repel Chinese and Japanese invasions and saw their country a battle-ground between these two rival powers. At different times each country has collected tribute, claiming Korea to be a vassal, and the state of vassalage had been admitted by the Korean monarchs. Manifestly such an admission of dependence upon each country in turn roused jealous feelings between China and Japan, each claiming dominance in the peninsula. Both countries laid claim to Korea, and both countries wasted armies and treasure through the centuries in the effort to maintain a semblance of authority or influence against the other. Sometimes a new Korean dynasty adopted a new policy toward the powerful neighbors, east and west, yielding tribute to one or both, or refusing it altogether, according to the condition of affairs within and without. Some of these wars were disastrous in the extreme, and the records of them which are preserved show battles of a ferocity and campaigns of a cruelty rarely excelled in history. In at least one Japanese invasion Christian missionaries participated, making many converts among the Koreans. Persecution of the adherents of the new faith followed, however, and Christianity was exterminated in Korea as effectually as it was in Japan.

Information in regard to Korea began to reach Europe in the Seventeenth Century. Jesuit missionaries in Peking sent home a map of the peninsula, and the Cossacks who overran Northern Asia brought reports of Korea to Russia, and from Russian sources came the first detailed information of the land. A Dutch ship was driven ashore in 1627 and the survivors, although kept as prisoners, gave to the Koreans what they knew of western arts and sciences. Thirty years later another vessel underwent the same experience, and after fifteen years of captivity these survivors escaped, and, returning to Holland, brought detailed information of what they had seen.

In 1777 Korean students in China gained some knowledge of Christianity, and returning to their home began to spread the new doctrines. The novel faith was welcomed, and converts were becoming numerous, when the Emperor issued an edict against Christianity, and a period of extermination began. Persecution and martyrdom marked the next few years, but the vital spark of Christianity remained among many people, and when more enlightened policies began to rule within recent years, missionaries found Korean families who had still preserved their faith for a century.

Within the last hundred years European and American vessels began to appear along the coast of Korea, in the effort to effect an entrance and obtain treaties. British, French, Russian and American fleets surveyed and mapped parts of the shores, but the policy of isolation remained impregnable, and there was no tolerance of communication. European missionaries who made their way into the country secretly were murdered, and it was impossible to obtain redress for the crimes. Nevertheless the Koreans began to feel a natural restlessness as they saw their neighbors gradually yielding to pressure from without, and permitting communication with what they considered the barbarian powers of Europe and America. The American squadron under Perry, and the treaty he made with Japan, gave the Koreans cause for thought. The treaty between China and Russia gave the Koreans a European power for their neighbor, immediately across the boundary northward. The French and English in 1860 opened war with China, took Peking, the capital, and drove the Chinese Emperor to flight, with a loss of prestige to China that struck terror to the hearts of the Koreans. With France and England on the west, Russia on the north, China humbled, and Japan open to the western world, it is not strange that the Korean rulers trembled.

It seemed to the rulers as if the world had conspired to break down their walls of isolation, and in the effort to beat back the inevitable, even more violent policies went into effect. Again the Christian converts and the missionaries who were laboring were assailed. Scores were put to death by torture, and hundreds were imprisoned. Instead of accomplishing the purpose desired, this was the final cause of world-entrance into Korea. French vessels patrolled the coast, entered rivers and destroyed native vessels and villages. Then a filibustering American vessel, endeavoring to effect a landing for purpose of trade or loot, as is variously contended, was captured and burned by the Koreans, and the crew and officers were killed. The United States steamship Shenandoah was sent to make an investigation, and came away convinced that the Americans had been the victims of their own folly in entering a land which was forbidden them, and in which the obligations of civilization and the white flag were not recognized.

However this may be, conditions along the Korean coast became intolerable, and another American squadron under Admiral Rodgers was sent to accomplish what Perry had done in Japan. The results were not as fortunate. In the unknown harbors vessels grounded, Korean forts bombarded the invaders, and finally, in a land battle, some 350 of the natives were killed, with an American loss of only thirteen in capturing the obsolete fortifications. Then the fleet sailed away, after thirty-five days in Korean waters, on July 3, 1871, with virtually no result accomplished.

This is but one of several such affairs witnessed along the Korean coast during the years from 1860 to 1880. The British and Japanese, among others, underwent similar experiences, and Russia, France and Italy all made futile efforts to negotiate treaties. The entering wedge was introduced by the Japanese, who induced the Koreans to send an embassy to Tokio, where they were received by the Mikado, and a treaty was arranged by which Fusan, on the south coast of Korea, long a Japanese outpost, was made a Japanese port for settlement and trade. The spirit of progress began to move even in Korea, and young men were sent to Japan and China to study and bring back what they could find that was good. In 1882 the Koreans yielded to the demand for treaties, and the United States, Great Britain, France, Germany, Italy and China all entered into conventions with the ruler of the peninsula.

Contact with western civilization has not yet done much to change the Korean manner of life and thought, nor to eliminate the disorders and jealousies in internal affairs. Indeed, these jealousies have multiplied since China, Russia and Japan all have been striving to obtain dominance over Korean affairs. Plots and counter plots have cursed the history of the little kingdom in the last decade, as truly as in the past. More than once the Emperor has fled from his palace to the protection of a foreign embassy, in order to save his life. Murders in the palace have been reported at intervals, and the actual authority of the government has been reduced to a minimum by the presence of powerful legation guards attached to each foreign ministry at the capital. Nevertheless, certain concessions for railway and telegraph building have been made, trade has expanded, missionaries have been permitted to labor, and there has been some advance in education.

The Koreans themselves are temperamentally of a gentle, orderly disposition, and in personal contact foreigners find them congenial as soon as a basis of acquaintance is established. They are frugal and industrious, as well as generous and hospitable, but their centuries of isolation and superstition have weakened their national vitality until in public affairs there is little to admire. The religion, art and literature are those of the Chinese, but of less vitality, and even the good things in the intellectual side of their life have fallen into a low state as a result of the general degeneracy and retrogression of the country.

We may take a brief glance at the war between China and Japan in 1894 and 1895, in order to lead up to the more important conflict of which this volume treats. That conflict was a result of the centuries of jealousy, divided authority and rival claims between Japan, China and Korea. Japan, in 1876, formally acknowledged Korean independence, but this acknowledgment was never imitated by China, which power always claimed suzerainty over the Hermit Kingdom. Korea had drawn her civilization from China, and the Chinese felt a parenthood for the little kingdom that they were not willing to relinquish. Japan, on the other hand, had conquered Korea in repeated invasions, had settled colonies of merchants and soldiers in two Korean ports, Fusan on the south, and Gensan, or Wonsan, as it is variously called, in Broughton Bay, a few hundred miles north on the eastern coast of the peninsula.

It would be confusing to attempt an analysis of the rival claims of Korean parties during the period of disorder that preceded the Chino-Japanese War. There were pro-Chinese and pro-Japanese elements, liberals and conservatives, loyalists and rebels, in inextricable confusion, except to those who study in detail the entire history of the peninsula and its politics. Riots and massacres occurred at intervals, during which leaders of the different parties fled for shelter to the foreign legations in the capital. Embassies from China, Japan and Korea dealt with the diplomats of Europe and America in support of their various claims. Assassination became a common weapon when diplomacy failed to accomplish purposes in Korea. Plots to murder the King and his ministers, or to blow up the government buildings, were reported at various times.

At last, Japan, as the nearest neighbor and the next friend, took the position that her own peace and welfare depended upon the reestablishment of order in Korea, under Japanese direction, to insure that it would be maintained. A scheme of reforms was drafted and submitted to the Koreans, with a proposal that China should join Japan to effect the desired objects. China refused even to discuss this proposal as long as any Japanese troops remained in

Korea, and Japan insisting that these were essential to preserve order, refused to withdraw. This was the signal for war, and fighting began at once.

The story of the operations of Chinese and Japanese forces during this war is but a record of continuous Japanese success. The utter inefficiency of the Chinese navy and the Chinese army was shown, and the effectiveness of preparations made by Japan was made equally manifest. The Japanese were the victors in every engagement on land and sea. When the war ended, the Chinese navy was destroyed or captured, and the Chinese army demoralized. The Japanese forces on land moved northward to Korea, as they have done in the early days of the Russo-Japanese War, sweeping the country and driving back Chinese forces wherever they were found. The most important and significant land engagements were those involved in the taking of Port Arthur, which was at that time a fortified Chinese military and naval station. The Japanese armies swept all before them, and in a masterly campaign, by most approved military methods, finally took the city by assault.

The naval operations were equally effective. The most important battle was that at the mouth of the Yalu River, the place of interest in the present war, as marking the line of separation between the Russians and Japanese during the early spring campaigning. In this naval victory many Chinese ships were destroyed and hundreds of sailors were killed or drowned. This battle was of the utmost interest to military students throughout the world, because of the fact that it was the first great fight of iron-clad vessels of modern construction, and it therefore afforded examples of high value as to the manner in which such craft would operate in actual engagements.

So invariable were Japanese success and Chinese defeat, that the end of the war soon came into sight. Early in 1895, with the Japanese in possession of Korea, Port Arthur and Wei-hai-wei, and with all the coast cities around the Yellow Sea subject to Japanese control, the defeated Chinese asked for an armistice and peace.

Protracted negotiations were carried on in Japan, the eminent Chinese statesman Li Hung Chang heading the Chinese Embassy to the Mikado.

In an earlier chapter we have already observed the terms of peace that were negotiated in the treaty, and the manner in which Russia interfered to deny Japan some of the fruits of victory. Since that time, as has been indicated, jealousies between Japan and Russia have multiplied, China has been in controversy with the world, in the "Boxer" outbreaks, and has seen her territory thrown open to trade and foreign settlement as a result, and Japan to-day claims in large degree the credit for the birth of western civilization into the Orient, and the increased prosperity that has come to the Koreans as well as to themselves and the Chinese.

The worship of ancestors is observed in Korea with as much punctiliousness as in China, but otherwise religion holds a low place in the kingdom. The law forbidding temples and priests in the city of Seoul has been repealed and all religions are permitted to teach their faith. Numerous Buddhist monasteries are scattered through the country. Confucianism is held in highest esteem by the upper classes, and a knowledge of the Chinese classics is the first aim of Korean scholars and aspirants for official station.

Seoul, the capital, with a population of about 200,000, contains an English church mission, with a bishop, and two hospitals attached to the mission, with trained nurses. The American missionaries have also two hospitals in Seoul. Altogether in the country there are about one hundred Protestant missionaries, and about one-third as many Roman Catholics.

In Seoul there is a school for English, with two English teachers and one hundred pupils. There are besides schools for teaching Japanese, French, German, Chinese and Russian, and a number of schools for little boys where Chinese and Korean are taught. All of these schools are subsidized by the government. The native language is intermediate between Mongolo-Tatar and Japanese, and an alphabetical system of writing is used to some extent. In all official writing and in the correspondence of the upper classes, the Chinese characters are used exclusively, but in official documents a mixture of the native script is the rule.

Korea is a purely agricultural country, and the methods of cultivation are of a backward and primitive type, the means of communication being few and difficult. In the south, rice, wheat, beans and grain of all kinds are grown, besides tobacco. In the north, the chief crops are barley, millet and oats. Rice, beans and ginseng are exported in large quantities. Gold, copper, iron and coal abound, and an American company is working gold mines in the mountains north of the capital. Similar concessions have been granted to a Russian company and to a German company.

Transport in the interior is entirely by porters, pack-horses and oxen. Small river steamers owned by Japanese run on the Han River, between Chemulpo, the port, and Seoul, the capital. A railway along the river has been built for the same distance by an American syndicate, and the Japanese, who already maintain a telegraph line between Seoul and Fusan on the south coast, are building a railway along that route. Other railway concessions have been granted, and the next few years are certain to see an immense improvement in the transportation facilities and a development of commerce, whatever may be the result of the war.

We have already seen something of the political relations between Russia and Manchuria, and the diplomatic methods by which Manchuria became to all intents and purposes a territory under Russian authority, with a Russian railway traversing it. Let us now glance at the Manchurian province itself, as regards its relations to the rest of the Chinese Empire, of which it has been so long a part, in its geography, its climate and its products.

Manchuria covers the northeastern part of the Chinese Empire. From Vladivostok, by a great circuit first northeast, then northwest and finally southwest to the point at which the Siberian Railway crosses into Manchuria, the province is virtually circumscribed by Russian territory. To the westward, however, it is bounded by the other provinces of the Chinese Empire, and on the south, Korea and the Yellow Sea mark its limits.

The province has an area of nearly 300,000 square miles, or about the same as that of Japan. Northern Manchuria is mainly mountainous, with a great range in the west, and others of less magnitude intersecting the province in different directions. There are extinct volcanoes, and much rough country outside of the actually mountainous region. Southern Manchuria is known as the Liao-tung peninsula, and forms a great triangle projecting into the Yellow Sea. This is the scene of the earlier war operations, and the entire region except the coasts is covered with low mountains whose summits do not exceed a height of 3,000 to 5,000 feet.

The mineral wealth of Manchuria is not yet fully explored, but according to the best information available it may be regarded as promising in the extreme. Coal beds occur at many places along the line of the railway, and along the coast of the Yellow Sea. Iron, silver, tin and gold are found in the mountain ranges, and rich gold deposits have been discovered quite near to Port Arthur.

The river systems of Manchuria are important in transportation, as well as in the development of the country. Along the entire northern boundary flows the great Amur River, which marks the frontier of Russia and China, and is navigable throughout its entire length. The Sungari, a tributary of the Amur, flows entirely across the center of the province, and the Ussuri forms the eastern boundary, between Manchuria and the Siberian province named for the river. Both of these streams are navigable by large river steamboats. Thanks to the Sungari, railway construction of the Manchurian line was made comparatively easy. Materials and laborers could be taken into the heart of the country by water with the utmost ease, so that construction was in progress from several points at the same time.

Ample rainfalls and frequent rivers and lakes make the province well-watered, generally speaking, and whatever agricultural resources it has in the soil, can be readily developed by the abundance of water.

The climate of the province varies greatly in its northern and southern portions. In northern Manchuria the winters are particularly severe, and much colder than any other parts of the world lying within the same latitudes. Winter lasts for five months, and the ice on the rivers attains a thickness of three or four feet. The spring is short and the change from cold to warm weather is very sudden, being accompanied by an extraordinarily rapid growth of vegetation. The greater part of the crop is sown in April. The summer is very warm, the autumn commences early, the leaves fall at the beginning of September, and morning frosts begin late in the same month. In the southern part of the province, however, the climate is modified by the warm currents from the Pacific, and the seasons are all milder.

The animal life and vegetation of northern Manchuria are much the same as those of Siberia, while southern Manchuria is more akin to China, Mongolia, Korea and Japan. A border province as it is, Manchuria includes not only the sable of the far north, but the Bengal tiger of southern Asia.

There are no very satisfactory figures of population, but the best estimates credit Manchuria with about 15,000,000 inhabitants. The mixture of races is remarkable, but the larger elements are the Chinese, Manchus, Koreans and Buriats. During the last century the Chinese have immigrated into the province in great numbers. Owing to the energy and industry of the Chinese, the central and southern portions of Manchuria differ but slightly in industry and manners of life from the parts of the Empire around Peking.

The chief occupation of the people of Manchuria is agriculture. Of late years the Chinese government has paid special care to agricultural development, inducing the settlement of unoccupied lands and the extensive cultivation of the more valuable crops. Wheat, oats, barley, millet, buckwheat, corn and rice are cultivated in the south. Poppies, from which opium is prepared, are an important product. Tobacco and cotton are raised, and large quantities of silk are produced. In the south, orchards and vineyards are important sources of wealth, and vegetables are grown in gardens everywhere. Apples, pears, peaches, plums, onions, red peppers and garlic are commonly found. The ginseng root is the most important medicinal plant in the opinion of the Chinese, and is grown in large quantities. Of late years this has been introduced into the United States from China, and is being grown to some extent for export to that country. Cattle raising is an important industry in Manchuria, and the horses of the provinces are famous. Camels are commonly employed for pack-animals.

Along the larger rivers there are fisheries of some importance, and on the sea coast these become a noteworthy industry, providing food not only for local consumption, but for export. Oysters also are found in the bays in the southern part of the province. Manufacturing industries are not at all developed, except in the form of handcraft to supply the needs of the local population.

During the few years immediately prior to the outbreak of the Russo-Japanese War, a period which measures the time of the Russian occupation of Manchuria, conditions in this province have materially changed from what was in effect before the Slavs came down from the North. It is the invariable policy of the Russians, wherever they obtain a foothold by conquest or by treaty, to subdue the native population and trade to their own demands, according to what they consider will be most profitable commercially, and most advantageous politically. In Manchuria this meant to garrison the line of the railway, to build Russian towns at strategic points, to police the province throughout, to establish branches of the Russo-Chinese Bank in every commercial town, and to encourage Russian merchants by favorable terms which could be given them for their freight shipments on the railway, their credits in the bank, and local contracts for the supply of provisions and materials for Russian military and engineering parties in the province.

All these things have been done as a part of the invariable Russian policy in conquered territory. The result has been the influx of a large Russian population in civil life. The Russian element in Manchuria, therefore, has not been measured entirely by the number of soldiers under arms, or the railway construction forces. Russian peasants have come to Manchuria in considerable number. Russian merchants and artisans have been everywhere. In this manner the Russification of Manchuria has advanced in every direction.

Of course, in large degree this Russian population has been concentrated in the larger towns of Manchuria. Temptations were few to attract isolated Russian settlers into Chinese communities. Their surroundings would not be congenial, nor would they have fair opportunities to prosper in their occupations. Instead they have centered in two or three of the more important Chinese cities of the province, forming large Russian colonies there, and in two instances, at least, new cities have been built that are virtually Russian in plan and population, except for the element of Chinese labor that has drifted to them.

One of these towns is Dalny, a few miles northeast of Port Arthur, on the southern coast of the peninsula which so early became the seat of war. When the Russians withdrew Port Arthur from the originally established privilege of a free port for all the nations, they created Dalny on the neighboring bay of Talienwan as a commercial substitute, in order that they might leave Port Arthur free to be fortified and maintained as a naval station without being subject to the observation of inquisitive foreigners. Dalny was hastened to completion in order to mollify the protesting nations, and rapidly became a real Russian town. But lightly fortified as it was, and conveniently prepared with docks, warehouses and shipping facilities, it was early made the point of Japanese attack and soon fell into the hands of the Islanders, to be used as a convenient station for their own operations.

The other of these purely Russian towns is Harbin, in the very heart of Manchuria, the junction point of the diverging lines of railway which traverse the province. From Harbin branches of the Chinese Eastern Railway, a part of the Siberian system, run to Mukden, Newchwang, Dalny and Port Arthur on the south, and to Nikolsk and Vladivostok on the east. From here also the main line of the Trans-Continental system runs northwestward to the Siberian boundary, connecting by way of Irkutsk, Omsk, and Cheliabinsk, with the Russian railways of the European continent. Harbin was but a village when the Russians came, although it was only a few miles from the important Manchurian city of Tsitsikhar. It is located at a point convenient to river navigation on the Sungari, and this combination of rail and river facilities suggested the location as the center of Russian settlement in Manchuria. A Russian city of considerable size has been built there, and with the Chinese population which has been drawn to the same place, the community is said to contain a population of nearly 70,000.

It is along this line of railway extending northward from Port Arthur to Harbin, that the Japanese have forced their advance, and the Russians have been compelled to fall back. In the process the Chinese inhabitants along the way, who, after all, have the best equity to the country, have been bitter sufferers. They have seen their fields devastated, their farming operations for the entire summer of 1904 interrupted, and their peaceful village life destroyed. The Russians have never been too lenient with the Asiatic races with whom they have come in contact, and in time of war it is not their disposition to waste energy on helpless people who get in the way. The campaign, therefore, has resulted in aggravating Chinese irritation against the Russians, who were from the beginning tolerated only because there was nothing else to do. In the course of several years of occupation of an Asiatic province, the Russians have usually been able to get on pretty good terms with the conquered tribes, but the Manchurian occupation has been too brief to establish this condition. The feeling of the natives to the Russian armies is, therefore, an important factor in the progress of events as the war proceeds, and will be one of the most interesting influences to watch with the advance of time, after the restoration of peace makes a readjustment of the entire situation necessary.

CHAPTER VI

UNITED STATES SAVES CHINA

Secretary of State John Hay Sends a Note to European Powers Asking Them to Join With the United States in Maintaining the Neutrality and Integrity of the Chinese Empire—All Accept and the "New Diplomacy" Wins a Signal Victory —Our Part in the War.

FROM the very beginning the United States has played an important part in the Russo-Japanese war. As soon as the two nations broke off diplomatic relations with each other Secretary of State John Hay cabled the following instructions to American ambassadors and ministers at the courts of the principal European governments:

"You will express to the minister of foreign affairs the earnest desire of the government of the United States that in the course of the military operations which have been begun between Russia and Japan the neutrality of China and in all practicable ways her administrative entity shall be respected by both parties and that the area of hostilities shall be localized and limited as much as possible, so that undue excitement and disturbance of the Chinese people may be prevented and the least possible loss to the commerce and intercourse of the world will be occasioned.

JOHN HAY."

This note marked a new departure in diplomacy, and was at first a surprise and puzzle to the European diplomats. Nearly all of the foreign governments delayed making a reply until they learned its exact import. It seemed rather audacious to tell two strong powers like Russia and Japan that they must keep their hands off China and must limit the area in which they fought to their own countries and the contiguous territory which had been the cause of the dispute. One by one, however, the other nations dropped in behind Uncle Sam, although some of them, like France, the ally of Russia, did so with great reluctance. The first ones, of course, were those that had large trade interests in China which would have been seriously disturbed, if not absolutely ruined, had the warring armies invaded the Chinese empire proper.

By the agreement of the powers brought about by Secretary Hay the trade of China was protected against the ravages of war, and that nation itself held in check, for Secretary Hay's note not only meant that Russia and Japan should respect the integrity of China, but that China herself should remain neutral and take no part in the war.

Secretary Hay's proposal was made for the purpose of keeping China out of the theater of war and without prejudice to Russia. The United States government felt that no nation could object to the suggestion that China remain neutral, and that no nation could justify itself if it did object.

In the first place, if China remained neutral the chances of China maintaining its integrity would be enhanced.

In the second place, Russia avoided the risk of having on her hands a war with China and was thereby left free to withdraw a large guard from her long frontier.

In the third place, the other nations, all of which had interests in China, had escaped the constant danger which would otherwise have been threatening them of becoming involved in the controversy.

The Two Main Propositions.

Secretary Hay's suggestion laid down the two propositions in the plainest language. He did not differentiate between old China and that part of the empire known as Manchuria and then occupied by Russia. Nor did he include the future independence of Korea. That was a question with which he did not want to involve the greater and more pressing question of the neutrality of China and the territorial integrity of China, which was a necessary corollary.

Courtesy of Everybody's Magazine.

JAPANESE FIRST CLASS BATTLESHIP YASHIMA.

Courtesy of Everybody's Magazine.

JAPANESE TORPEDO BOAT DESTROYER USUGUMA.
One of the Swift Vessels of the Japanese Navy.

Nothing that has happened in American diplomacy since Richard Olney, then American Secretary of State, sent his famous message in the Venezuelan boundary dispute that "the United States are practically sovereign on this continent and their fiat is law," has aroused as much comment as Secretary Hay's note with respect to China's neutrality and the preservation of her empire, although the latter was a subject upon which men of different political views held different opinions. Right or wrong, it was accomplished and China was given a new reason to be grateful to the United States of America.

Inspired By Germany.

Great Britain and Germany made ready responses to the American note expressing their approval. It afterward transpired that the note was inspired by Baron Speck Von Sternberg, the German Ambassador to the United States. It was also learned that the British Ambassador had been instructed to suggest to the American Secretary of State that he take the initiative in demanding that the neutrality of China be preserved, and the integrity of that country respected, but the German Ambassador was the first to have a conference with Secretary Hay on the subject.

The reason why the United States took the lead in this matter was that this country had no territorial interests in China, no "spheres of influence" like Great Britain, Germany and France, and hence was the only great power occupying an independent position in all matters relating to the Chinese empire. Had either of the European powers mentioned addressed such a note to the two belligerents they would have been open to the charge of self-interest, and certainly their motives would not have been above suspicion.

Even as it was, Secretary Hay was accused in some quarters of playing into the hands of Japan, and while the note caused no friction with the Russian government, it aroused a great deal of indignation among the Russian people, while the Russian press denounced Secretary Hay and the United States as enemies of the Czar's government. The important replies to the note, of course, were those from the two warring powers. Inasmuch as Japan was fighting to maintain the integrity of the Chinese empire, the Mikado's government gave its acquiescence. The Japanese reply, which was communicated through the American Minister at Tokio, on February 13, was as follows:

"In response to your note of the 12th inst., on the subject of the neutrality of China during the existing war, I beg to say that the imperial government, sharing with the government of the United States in the fullest measure the desire to avoid as far as possible any disturbance of the orderly condition of affairs now prevailing in China, are prepared to respect the neutrality and administrative entity of China outside the regions occupied by Russia as long as Russia, making a similar engagement, fulfills, in good faith, the terms and conditions of such engagement."

Russia Exempts Manchuria.

Nearly a week later, or on February 19, the following reply was received from the Russian government:

"The imperial government shares completely the desire to insure tranquility of China; is ready to adhere to an understanding with other powers for the purpose of safeguarding the neutrality of that empire on the following conditions:

"1. China must herself strictly observe all the clauses of neutrality.

"2. The Japanese government must loyally observe the engagements entered into with the powers as well as the principles generally recognized by the law of nations.

"3. That it is well understood that neutralization in no case can be extended to Manchuria, the territory of which, by the force of events, will serve as field of miliatry operation."

It will be noted that Russia particularly exempted Manchuria from the neutral zone, and as that great province of China is the only section in which Russian interests would be affected, it was manifestly to the advantage of Russia that the remainder of China should be kept neutral.

Secretary Hay sent the following telegram to Russia and Japan, the import of which was communicated to the other powers interested:

"The answer of the Russian government is viewed as responsive to the proposal made by the United States, as well as by the other powers, and this government will have pleasure in communicating it forthwith to the governments of China and Japan, each of which has already informed us of its adherence to the principles set forth in our circular proposal."

China Promises Neutrality.

In the meantime, an exchange of correspondence took place between the governments of China and Japan, concerning the question of Chinese neutrality and Japan's attitude toward the Chinese empire.

In a note handed by the Chinese Minister at Tokio to the Japanese Minister for foreign affairs, February 13, China announced that the Pekin government had taken steps to observe the rules of neutrality between nations and had ordered all the authorities of the provinces to observe them strictly. Special preparations had been ordered for the guarding of Mukden and Shinkin, the sites of the imperial mausoleums and palaces. A garrison has been dispatched to all the districts west of the Liao which had been evacuated by Russia.

China engaged, however, not to take such steps "as will make a rupture of friendly relations in Manchuria," adding that "there are localities still in occupation by foreign troops and beyond the reach of the power of China, where the enforcement of such rules of neutrality, it is feared, will be impossible. The three eastern provinces, as well as the rights pertaining thereto, shall remain under China's sovereignty whatever side may gain the victory, and shall not be occupied by either of the powers now at war."

Japan's Promise to China.

The response of Japan to the request of China for a statement of its position in regard to the neutrality of the latter country is as follows:

"The imperial government, desiring to avoid as far as possible a disturbance of the peaceful condition of affairs which prevails in China, will in all parts of Chinese territory, excepting the regions now occupied by Russia, respect the neutrality of China so long as Russia does the same.

"The rules of war which govern the forces of Japan in the field do not permit the wanton destruction of property. Accordingly, the imperial Chinese government may rest assured that the mausoleums and palaces at Mukden and Hsing-Kiang and public buildings in China everywhere will be secure from any injury not attributable to the action of Russia.

"Furthermore, the rights of Chinese officials and inhabitants within the zone of military operations will in their persons and property be fully respected and protected by the imperial forces, so far as military necessity permits. In the event, however, that they should extend aid and comfort to the enemy of Japan, the imperial government reserves to itself the right to take such action as the circumstances require.

"It remains to say, in conclusion, that the war is not being waged by the Japanese for the purpose of conquest, but solely in defense of her legitimate interests, and, consequently, the imperial government has no intention to acquire territory at the expense of China as a result of the conflict.

"The imperial government also wishes the imperial Chinese government to clearly understand that whatever action may be taken by it on Chinese territory which is made the theater of war will be the result of military necessity and not impairment of Chinese sovereignty."

CHAPTER VII

DUTIES OF NEUTRAL NATIONS

President Roosevelt's Proclamation of Neutrality Defines the Obligations of a Neutral Power and of its Citizens or Subjects, and Also the Rights of the Belligerents With Respect to Neutral Nations—All the Great Powers Declare Neutrality at the Outbreak of the Russo-Japanese War.

ALL of the great powers of Europe, together with the United States, issued proclamations of neutrality, after Russia and Japan each had made formal declaration of war.

President Roosevelt's proclamaticn covers so thoroughly the duties and obligations of a neutral power and its citizens that it is herewith given in full:

"Whereas, A state of war unhappily exists between Japan on the one side and Russia on the other side; and,

"Whereas, The United States are on terms of friendship and amity with both the contending powers, and with the persons inhabiting their several dominions; and,

"Whereas, There are citizens of the United States residing within the territories or dominions of each of the said belligerents, and carrying on commerce, trade, or other business or pursuits therein, protected by the faith of treaties; and,

"Whereas, There are subjects of each of the said belligerents residing within the territory or jurisdiction of the United States, and carrying on commerce, trade, or other business or pursuits therein; and,

"Whereas, The laws of the United States, without interfering with the free expression of opinion and sympathy, or with the open manufacture or sale of arms or munitions of war, nevertheless impose upon all persons who may be within their territory and jurisdiction the duty of an impartial neutrality during the existence of the contest; and,

Will Preserve Neutrality.

"Whereas, It is the duty of a neutral government not to permit or suffer the making of its waters subservient to the purposes of war;

"Now, therefore, I, Theodore Roosevelt, President of the United States of America, in order to preserve the neutrality of the United States and of their citizens and of persons within their territory and jurisdiction, and to enforce their laws, and in order that all persons, being warned of the general tenor of the laws and treaties of the United States in this behalf and of the law of nations, may thus be prevented from an unintentional violation of the same, do hereby declare and proclaim that by the act passed on the 20th day of April, A. D. 1818, commonly known as the 'neutrality law,' the following acts are forbidden to be done, under severe penalties. within the territory and jurisdiction of the United States, to wit:

"1. Accepting and exercising a commission to serve either of the said belligerents by land or by sea against the other belligerent.

"2. Enlisting or entering into the service of either of the said belligerents as a soldier or as a marine or seaman on board of a vessel of war, letter of marque, or privateer.

"3. Hiring or retaining another person to enlist or enter himself in the service of either of the said belligerents as a soldier or as a marine or seaman on board of any vessel of war, letter of marque, or privateer.

"4. Hiring another person to go beyond the limits or jurisdiction of the United States with intent to be enlisted as aforesaid.

"5. Hiring another person to go beyond the limits of the United States with intent to be entered into service as aforesaid.

"6. Retaining another person to go beyond the limits of the United States with intent to be enlisted as aforesaid.

"7. Retaining another person to go beyond the limits of the United States to be entered into the service of the aforesaid. (But the said act is not to be construed to extend to a citizen of either belligerent, who, being transiently within the United States, shall, on board of another vessel of war which at the time of its arrival within the United States was filled and equipped as such vessel of war, enlist or enter himself or hire or retain another subject or citizen of the same belligerent who is transiently within the United States to enlist or enter himself to serve such belligerent on board such vessel of war if the United States shall then be at peace with such belligerent.)

Privateering Is Prohibited.

"8. Fitting out and arming, or attempting to fit out and arm or procuring to be fitted out and armed, or knowingly being concerned in the furnishing, fitting out or arming of any ship or vessel with intent that such ship or vessel shall be employed in the service of either of the belligerents.

"9. Issuing or delivering a commission within the territory or jurisdiction of the United States for a ship or vessel to the intent that it may be employed as aforesaid.

"10. Increasing or augmenting or procuring to be increased or augmented, or knowingly being concerned in increasing or augmenting the force of any ship of war, cruiser, or other armed vessel which at the time of its arrival within the United States was a ship of war, cruiser, or armed vessel in the service of either of the said belligerents, or belonging to the subjects of either, by adding to the number of guns of such vessels, or by changing those on board of it for guns of a larger caliber, or by the addition thereto of any equipment solely applicable to war.

"11. Beginning or setting on foot, or providing or preparing the means for any military expedition or enterprise to be carried on from the territory or jurisdiction of the United States against the territory or dominions of either of the said belligerents.

Closed to Hostile Ships.

"And I do hereby further declare and proclaim that any frequenting and use of the waters within the territorial jurisdiction of the United States by the armed vessels of either belligerents, whether public ships or privateers, for the purpose of preparing for hostile operations, or as posts of observation upon the ships of war or privateers or merchant vessels of the other belligerent lying within or being about to enter the jurisdiction of the United States, must be regarded as unfriendly and offensive and in violation of that neutrality which it is the determination of this government to observe.

"To the end that the hazard and inconvenience of such apprehended practices may be avoided, I further proclaim and declare that from and after the 15th day of February instant, and during the continuance of the present hostilities between Japan and Russia, no ship of war or privateer of either belligerent shall be permitted to make use of any port, harbor, roadstead, or waters subject to the jurisdiction of the United States from which a vessel of the other belligerent—whether the same shall be a ship of war, a privateer, or a merchant ship—shall have previously departed until after the expiration of at least twenty-four hours from the departure of such last mentioned beyond the jurisdiction of the United States.

May Stay Twenty-four Hours.

"If any ship of war or privateer of either belligerent shall, after the time of this notification, enter any port, harbor, roadstead, or waters of the United States such vessel shall be required to depart and put to sea within twenty-four hours after its entrance into such port, harbor, roadstead, or waters, except in case of stress of weather or of it requiring provisions or things necessary for the subsistence of the crew, or for repairs; in either of which cases the authorities of the port or of the nearest port (as the case may be) shall require it to put to sea as soon as possible after the expiration of such period of twenty-four hours without permitting it to take in supplies beyond what may be necessary for immediate use, and no such vessel which may have been permitted to remain within the waters of the United States for the purpose of repair shall continue within such port, harbor, roadstead, or waters for a longer period than twenty-four hours after necessary repairs shall have been completed unless within such twenty-four hours a vessel, whether ship of war, privateer; or merchant ship of the other belligerent shall have departed therefrom, in which case the time limited for the departure of such ship of war or privateer shall be extended so far as may be necessary to secure an interval of not less than twenty-four hours between such departure and that of any ship of war, privateer, or merchant ship of the other belligerent which may have previously quit the same port, harbor, roadstead, or waters.

For Harbor Regulations.

"No ship of war or privateer of either belligerent shall be detained in any port, harbor, roadstead, or waters of the United States more than twenty-four hours by reason of the successive departures from such port, harbor, roadstead, or waters of more than one vessel of the other belligerent. But if there be several vessels of each or either of the two belligerents in the same port, harbor, roadstead, or waters, the order of their departure therefrom shall be so arranged as to afford the opportunity of leaving alternately to the vessels of the respective belligerents, and to cause the least detention consistent with the objects of this proclamation.

"No ship of war or privateer of either belligerent shall be permitted while in any port, harbor, roadstead, or waters within the jurisdiction of the United States to take in any supplies except provisions and such other things as may be requisite for the subsistence of the crew, and except so much coal only as may be sufficient to carry such vessel, if without any sail power, to the nearest port of its own country, or in case the vessel is rigged to go under sail and may also be propelled by steam power, then with half the quantity of coal which it would be entitled to receive, if dependent upon steam alone, and no coal shall be again supplied to any such ship of war or privateer in the same or any other port, harbor, roadstead, or waters of the United States without special permission, until after the expiration of three months from the time when such coal may have been last supplied to it within the waters of the United States unless such ship of war or privateer shall, since last thus supplied, have entered a port of the government to which it belongs.

Rights of Neutral Ships.

"And I further declare and proclaim that by the first article of the convention as to rights of neutrals at sea, which was concluded between the United States of America and his majesty the Emperor of all the Russias, on July 22, A. D. 1854, the following principles were recognized as permanent and immutable—to wit:

"1. That free ships make free goods; that is to say, that the effects or goods belonging to subjects or citizens of a power or state at war are free from capture and confiscation when found on board neutral vessels, with the exception of articles of contraband of war.

"2. That the property of neutrals on board an enemy's vessel is not subject to confiscation, unless the same be contraband of war.

"And I do further declare and proclaim that the statutes of the United States and the law of nations alike require that no person, within the territory and jurisdiction of the United States, shall take part, directly or indirectly, in the said war, but shall remain at peace with each of the said belligerents, and shall maintain a strict and impartial neutrality, and that whatever privileges shall be accorded to one belligerent within the ports of the United States shall be in like manner accorded to the other.

"And I do hereby enjoin all the good citizens of the United

States and all persons residing or being within the territory or jurisdiction of the United States to observe the laws thereof, and to commit no act contrary to the provisions of the said statutes or in violation of the law of nations in that behalf.

Contraband of War.

"And I do hereby warn all citizens of the United States and all persons residing or being within their territory or jurisdiction that while the free and full expression of sympathies in public and private is not restricted by the laws of the United States, military forces in aid of either belligerent cannot lawfully be originated or organized within their jurisdiction, and that while all persons may lawfully, and without restriction by reason of the aforesaid state of war, manufacture and sell within the United States arms and munitions of war, and other articles ordinarily known as 'contraband of war,' yet they cannot carry such articles upon the high seas for the use or service of either belligerent, nor can they transport soldiers and officers of either, or attempt to break any blockade which may be lawfully established and maintained during the war, without incurring the risk of hostile capture, and the penalties denounced by the law of nations in that behalf.

"And I do hereby give notice that all citizens of the United States and others who may claim the protection of this government, who may misconduct themselves in the premises will do so at their peril and that they can in nowise obtain any protection from the government of the United States against the consequences of their conduct.

"In witness whereof I have hereunto set my hand and caused the seal of the United States to be affixed.

"Done at the city of Washington this 11th day of February, in the year (seal) of our Lord one thousand nine hundred and four and of the independence of the United States the one hundred and twenty-eighth. THEODORE ROOSEVELT.

"By the President.

"JOHN HAY, Secretary of State."

CHAPTER VIII

FIGHTING FORCES OF THE MIKADO AND THE CZAR

Comparison of the Military, Naval and Financial Strength of the Combatants at the Outbreak of the War—Japan Well Prepared in Every Way, but Her Army Was Small in Comparison With the Russian Military Machine—Financial Standing and Resources of the Two Warring Empires.

SOLDIERS, ships and money are the three fighting forces of a nation. A comparison of the financial and naval strength of Russia and Japan at the beginning of the war showed plainly that the little island nation was well prepared in these two particulars, although the strength of her army was small compared to the great military force of the Russian empire. At the same time it was apparent that if Japan could control the sea and suffer no reverses in her monetary resources that she could easily place on the Asiatic continent an army which Russia would be perplexed to crush. It had been repeatedly alleged in ill-informed circles that the state of Japan's finances handicapped her heavily for fighting purposes.

Precisely the same estimate found general credence at the outset of the war in 1894. But in truth Japan's position at the beginning of the war with Russia was incomparably better than it was then. She had in the vaults of the Central Bank specie aggregating 113,000,000 yen (£11,300,000), an altogether unprecedented amount. There also remained to the bank a legal margin of 35,000,000 of note-issuing power, which could be expanded to at least 50,000,000 when the invariable year's end drain was succeeded by the spring deposits. Out of the proceeds of the 50,000,000 yen worth of bonds sold abroad last year a great part lay in hard money in London. Her next fiscal year's budget showed a revenue of 225,225,000 yen, against an expenditure of 183,667,000, a surplus of over 41,000,000 being thus available.

The treasury had also in hand the three capital funds—the naval maintenance fund (30,000,000), the education fund (10,000,000) and famine relief fund (10,000,000)—one-half of this total of 50,000,000 being in specie. Finally, the market rate of interest, which ranged from 10¾ to 12¼ last year, was quoted at 6¼, and the banks throughout the country were embarrassed with funds for which they could not find any profitable use.

Japan's actual outlays during her war with China in 1894-5 were 165,225,000 yen on account of the army and 35,000,000 yen on account of the navy, in which figures there were included large sums for the purchase of transports, men-of-war and ammunition, none of which expenses would have to be incurred in her war with Russia. As for the hard money side of the account, the aggregate outlays did not reach 12,000,000 yen. It may be said, therefore, that the financial situation was notably favorable for Japan.

Sea Forces Nearly Equal.

Concerning the command of the sea, the latest statement from authoritative Japanese sources showed the strength of the two squadrons as follows, premising that the Russian totals included vessels then en route for the far east: Of battle-ships Japan had six, with a displacement of 86,299 tons, against nine Russian, with a displacement of 110,232 tons, but this superiority on Russia's side was somewhat redressed by the greater size, higher speed and heavier armament of the Japanese ships.

Of armored cruisers Japan had six, with a displacement of 58,788 tons, against five Russian, with a displacement of 49,016 tons.

Of protected cruisers capable of developing a speed of twenty knots and upward Japan had six, with a displacement of 25,106 tons, against eight Russian, with a displacement of 45,553 tons. In this list, then, which comprised first-class fighting material,

Japan's ships numbered eighteen, or a total tonnage of 170,182, against twenty-two Russian, with a total tonnage of 204,801.

In second-class fighting material Japan had twelve cruisers of a speed of sixteen knots and upward, with a displacement of 37,739 tons, against one Russian, with a displacement of 3,285 tons, and Japan had twenty-three gunboats and coast defense ships, aggregating 28,391 tons, against twelve Russian, aggregating 12,988 tons. Thus, in second-class material Japan had thirty-five ships, representing 66,130 tons, against thirteen Russian, totaling 26,273 tons.

Of torpedo destroyers Japan had nineteen, displacing 6,227 tons, against thirty-two Russian, displacing 9,608 tons, and of torpedo boats Japan had eighty-five, against Russia's fourteen. On the whole, then, the fighting forces of the two powers seem tolerably equal, while Japan had a great advantage in the proximity of her base and in the possession of ample docking facilities, the latter a point where Russia was probably inferior.

Russia's Enormous Army.

Though there was little possibility that the Russians would put half or even a third of their effective war strength into the field, and though it had been announced that no troops would be withdrawn from the Austrian or Bessarabian frontiers, it may be interesting to state what the actual war strength of the Russian army was. It should be remembered that every year 890,000 young men are liable to conscription, of whom 220,000 are employed, the remainder passing into the reserve. The total war strength of the active army was twenty-four army corps, with fifty-one infantry and twenty-three cavalry divisions, one rifle division, two rifle brigades. The reserve consists of twenty divisions of the first and fifteen divisions of the second class, 490 Cossack squadrons and seventeen horse artillery batteries of six fifteen-pounders each. The active army consisted of 989 battalions of infantry, 730 squadrons of cavalry, 3,782 guns, 161 engineer companies, 29,000 officers and 1,266,700 men. The reserves, including local and fortress troops, brought the strength to 2,870 battalions, 1,600 squadrons, 6,500 guns, 75,000 officers and 4,500,000. These figures are in round numbers.

The peace strength of a regiment is seventy officers and 1,887 men, twenty-five horses and no wheeled vehicles. In war the total is raised.

The Russian uniform is green, with a dark green flat cloth cap. The great coat is gray, reaching half way between knee and ankle. In summer white linen blouse and trousers are worn. Equipment consists for each man of three cartridge pouches and a bandolier, carrying 120 rounds in all; a kit bag, an aluminum water bottle, a great coat and the sixth part of a shelter tent; total weight, 58.22 pounds. This includes an emergency ration of biscuit and salt. Meat is carried "on the hoof"; cattle for slaughter accompany the marching columns.

Soldiers Must Pray.

Every morning and every night the Russian soldier is summoned to prayers. The services are as much a part of the everyday routine as breakfast and supper. No other army observes so many religious ceremonies.

With drilling and riding, gymnastics, fencing and shooting, according to his regiment, the soldier works hard until the time for dinner arrives, between 11 and 12. Afterward, until 2 o'clock, he may sleep or rest. Two hours' drilling is followed by tea. Between 6 and 7 the illiterates of the regiment study the arts of reading and writing in large classes.

On this ocasion the Czar rides all round the camp, which is several miles in circumference, and inspects each regiment. The troops are without their side arms, for this is largely a religious ceremony. As the Czar passes he greets the soldiers with the well-known words, "Zdorovo molotzee"—"Your health, my lads"—which they answer with, "Your imperial majesty's good health," and with loud hurrahs.

The Czar, having completed his inspection, returns to the imperial marquee, pitched on the top of a hill, and receives the reports of the officers. Then at a signal all the guns around the camp, numbering some 500 pieces, salute, while a thousand musicians and drummers strike up the "Kolj Slaven," a Russian national hymn. At another signal all the soldiers present begin slowly to intone the Lord's Prayer in unison. The effect is indescribably impressive.

Since the Turkish war the Russian army has been thoroughly reorganized by General Vanoffsky. When supplied with the new quick-firing guns now rapidly being introduced, next to Germany, it will be the finest army in the world. The Turkish war was an eye-opener to the Russian government as regards the inefficiency of the army. Had it not been for the corruption of the Turkish pashas the Russian army in Turkey would never have gone home again.

An idea of the efficiency of the Russian army may be gained from the recent Russian conquest of Manchuria, which was effected almost entirely by Siberian troops, the troops at home having been left practically untouched.

In this campaign Russia swiftly and secretly massed 150,000 efficient troops on the Manchurian frontier, thousands of miles from Europe, without any special effort.

The Fighting Cossacks.

The Cossacks, which form a branch of the Russian army service which has no parallel in any other army in the world, are irregular cavalry, but very different from volunteers of America, for the right to be a Cossack comes only through inheritance. The son of a Cossack, therefore, is a Cossack as soon as he is born, and is taught the use of arms and the traditions of his warrior race all through his boyhood.

So exclusive is this hereditary military caste that it is almost impossible for an officer of the regular army, no matter how

JAPANESE JINRICKSHAWS.
A Typical Scene in Every Japanese City—How Native Women Are Drawn About by Coolies.

JAPANESE IDOLS IN THE TEMPLE OF NIKKO.
Effigies Worshipped by Japanese.

high his rank, to secure a commission in a Cossack regiment unless he has inherited the right to such a command.

Several hundred years ago the Cossacks were lawless bands of freebooters living on the banks of the Don and Dneiper in Russia. When this territory was absorbed into the Russian dominions the Czars were confronted with the problem of turning these turbulent people into good citizens. They had always been trained to martial life and the use of arms, so the most natural and the best solution of the problem seemed to be to turn them into soldiers. The experiment proved a great success and ever since it was put into execution the Cossacks have been of the greatest assistance to Russia in all her military enterprises, and to-day there is no more familiar name in connection with the Russian army than that of "Cossack."

They are organized into regiments, but it is only certain of the officers who are on duty all the time. Each man gets his horse and a small pay from the government when not on active duty, and is allowed to settle down and rear a family of young Cossacks for the service of the Czar.

Fighting Force of Japan.

Compared with the huge military strength of Russia, that of Japan was little more than a pygmy one. Yet it reached the fairly substantial total of about 450,000 from all sources, together with 1,200 guns and 90,000 horses.

The fighting force always available for duty was divided into 156 batallions of infantry, fifty-one squadrons of cavalry, forty batteries of field and fortress artillery, together with a proportionate complement of engineer, supply and transport detachments. For armament the field artillery had the Arisaka quick-firing pattern, and the infantry on the permanent establishment carried the Midji magazine rifle. Japanese cavalry are served out with swords and carbines, but no lances.

The Japanese navy had its beginning in 1866, when Japan purchased the armor-clad Stonewall Jackson (afterward named the Adzuma), and under American instructors, they have gone on steadily increasing their armored fleet, in addition to building up an unarmored fleet, all armed with the best rifled guns. The first armored ship constructed for Japan was built on the Thames and was launched in 1877, about six years before our new navy was begun. She was the Foo-So, and had a displacement of 3,718 tons. About the same time contracts were made in England for the two composite armor-belted corvettes, the Kon-Go and the Hi-Yei. Then, in 1885, the Naniwa and the Takachiho, built by the Armstrongs, in England, were launched. They were protected cruisers of 3,700 tons displacement and eighteen knots speed, and were conspicuous in the Japan-Chinese war.

At the outbreak of the war, Japan ranked seventh in naval strength among the powers of the world. The efficiency of the navy is due to the training received under American naval officers, particularly H. Walton Grinnell, who, in 1868, accepted a commission of admiral of the Japanese navy and served three years.

CHAPTER IX

GREAT MEN OF THE WAR

Brief Sketches of the Men who Rule the Destinies of Japan and Russia and Others Famous in Diplomacy, the Army and the Navy—Pen Pictures of Mutsuhito the Japanese Mikado, Viceroy Alexieff, Admiral Togo, Plehve the "Terrible Russ," &c., &c., Together With Anecdotes Illustrating Their Chief Characteristics.

COMPARATIVELY few foreigners have seen the Mikado of Japan closely. In spite of its wonderful advance in Occidental ideas in recent years, Japan retains enough of its Orientalism to insist upon a certain seclusion for its ruler. Mutsuhito breaks away from his purely Oriental environment occasionally. He goes among his people incognito. While strolling through the streets of Tokio as a young man attired as a common Japanese sailor, Mutsuhito encountered the first American he had ever seen. Walking boldly up to the son of "Uncle Sam," the boy Emperor introduced himself as a young sailor, and, finding the American could speak a little Japanese, he poured forth a flood of eager questions. The traveler from the United States told the supposed sailor a wonderful tale of the results of American civilization. The imperial ambition received new stimulus, and that interview with an American accomplished much for Japan.

Mutsuhito-Tenwo, Emperor of Japan, is the present representative of the oldest royal dynasty extant. He is the one hundred and twenty-first Emperor of his dynasty, which dates back in an unbroken line 2,555 years. He is the direct descendant of Ginmu, the "Divine Conqueror," who, according to Japanese mythology, "descended from heaven on the white bird of the clouds." Ginmu's first task in his mythological role of divine conqueror was the subjugation of the Ainos, a savage, warlike race, whose descendants are still found in the northern extremity of Japan. Having subdued these fierce Ainos, Ginmu proclaimed himself to be "Tenshi," the "Son of Heaven," and established the still existing dynasty in 660 B. C. It is no exaggeration, therefore, to say that through the veins of Mutsuhito-Tenwo flows the very bluest of "blue blood."

Personally, the Emperor has a pleasant appearance. He is very tall for a Japanese, almost six feet. He is muscular and well proportioned. He has a broad, high forehead, and, judged by the most exacting standard of manly beauty, is a handsome sovereign.

The Mikado takes more interest in the government than any of his predecessors. He reads the papers and attends Cabinet councils. He takes all the important American and English magazines. He has astonished the upper classes of Japan by knowing something about the government of his people.

The Mikado lives in a palace built in the American way, with steel framework made in Pittsburg, Pa. This was done to avoid accidents by earthquakes, so common in Japan.

Haruko, Empress of Japan, was a daughter of a Japanese noble. She is 54 years old, two years older than her husband. Her name, Haruko, means "spring time."

In the Mikado's reign the bands of feudalism that bound Japan to the Middle Ages were broken; a constitution was granted by him volutarily; the old social order of caste limitations gave way to a more liberal order of equality; modern education, literature, arts, science and industry were welcomed; the army and navy were changed from the bow and arrow stage to modern organizations. It was only this remarkable advancement in the reign of Mutsuhito that made it possible for Oriental Japan to be equal to the task of a possible successful war with Russia.

Admiral Togo.

Admiral Togo, the man who commanded Japan's splendid fleet in the attack upon Port Arthur, is about 55 years old. He is not of princely or noble birth, but is a simple gentleman, a Samurai of the great Satsuma clan, as so many of his fellow officers are.

He received a great part of his education at the English naval college at Greenwich.

Returning to Japan, he was employed in various capacities. In 1894, when war broke out between China and Japan, he was in command of the Naniwa, a cruiser of 3,650 tons, one of the vessels composing what was then known as the first flying squadron of the Japanese navy. During the war he greatly distinguished himself and earned the reputation of being a first-class fighting man.

It was in connection with the episode of the sinking of the Kowshing that his name first came into world-wide prominence. At the time some adverse criticisms of his action on that occasion were heard, but these died away on a fuller knowledge of the circumstances. The story reveals something of the character of the man, so it is worth telling again.

The Kowshing, a transport vessel flying the British flag, with a British captain and crew, and carrying some 1,100 Chinese soldiers for Asan, was met by Togo in the Naniwa, who signaled her to stop. A Japanese lieutenant went on board with a peremptory order from Togo that the transport must proceed no further toward her destination, but at once accompany the Naniwa to the main Japanese fleet.

Captain Galsworthy of the Kowshing was willing to obey these orders, but not so minded were the officers of the Chinese forces on the vessel; they immediately raised a great clamor and threatened Galsworthy.

Seeing what was occurring Togo sent a boat to bring off Captain Galsworthy and his crew, but the Chinese prevented them from leaving. Finally Togo signaled Galsworthy to take one of his own boats and come over to the Naniwa, but the British captain was not allowed by the Chinese to do so.

For four hours Togo stood off, in an effort to save Galsworthy and the ship. Then he hoisted the red flag, which announced that he was about to open fire. A few moments later a well-directed shot from the Naniwa struck the engine-room and penetrated the hull of the Kowshing, which soon afterward filled and sank.

As Galsworthy and his men leaped over the bulwarks of the transport into the sea they were fired on by the Chinese. Togo at once sent out boats and rescued as many as he could.

In this way Togo began the Chino-Japanese war. His countrymen have never forgotten the part he played in this episode. "Togo!" they say, "it was Togo who sank the Kowshing." And they draw a confident augury from it.

Plehve, the Terrible.

The most powerful man in all Russia without any doubt is the minister of the interior, Secretary of State and Senator Wjatscheslavo Konstantinovitsch Von Plehve, who was intrusted with the management of Russian home affairs in April, 1902, after the murder of Minister Syingin.

Plehve was born in 1848, the son of a poor nobleman. An aristocrat of Polish descent patronized him and gave him an excellent education. He studied law in Moscow and was appointed assistant to the procurator of the law courts of Moscow.

He was soon recognized as exceptionally clever and in a very short time made rapid strides in his career, occupying the posts of procurator in Vladimir, Tula and Vologa.

But it was as imperial counsel at the courts of Warsaw that he first really distinguished himself. He put all considerations aside and devoted his great mind to the persecuting of the Polish elements dissatisfied with the oppressive regime of the Russian government. He went so far as to involve the family which had benefited him in his youth in a disastrous law suit, merely to prove his earnest desire to please the Russian government.

He discovered so many instances of high treason in Poland that government attention was called to him, and he was named procurator of the courts of St. Petersburg. Among his duties was to make a daily report to Czar Alexander II on the examination of all nihilistic crimes.

Plehve showed unparalleled energy in discovering and arresting the criminals who tried to blow up the Winter Palace. Then he invented moral torments, by means of which he extorted confessions.

After Czar Alexander III was assassinated Plehve was named first chief of the state police and president of the commission instructed to find means for restoring order in the empire. In this double position Plehve exercised a terrorism not heard of even in Russia. He sent thousands of intelligent Russian citizens to Siberia or put them in prison for life.

This reign of terror lasted three years, when, in 1884, Plehve was appointed assistant to the minister of the interior. He took an active part in all the reactionary measures of the reign of Alexander III. He did all in his power to annihilate German colonists in Russia, the nobility in the Baltic provinces and the Jews. It seems to be his aim to extirpate all elements which are not orthodox Russian.

He devised the institution of class presidents, authorized to terrorize the people.

The accession of Czar Nicholas again brought him advancement. In 1901 he was named state secretary for Finland, where he nearly caused a revolt by his cruel measures.

After he undertook the management of home affairs, there were an uninterrupted series of riots at the universities and among the peasants of the provinces of Charkoff, Poltava, Cherson, Tschanigoff, Saratoff and Kieff. He banished untold numbers of students to Siberia.

This was the policy which developed events such as they have happened at Kishineff, and will undoubtedly happen elsewhere. Plehve is an open enemy of the Jews, whose lawful rights he denies and refuses to protect.

Yamagata, Soldier and Statesman.

Next to Marquis Ito, the greatest figure in Japan is Field Marshal Yamagata, the Japanese von Moltke, soldier and statesman, veteran of many wars and former prime minister of the empire.

Yamagata is not only the first in rank in the Japanese army but he is also first in the esteem of the Japanese people. He has been fighting the battles of Japan since 1868, when he took part in the struggle that restored the Emperor to power.

In the following year he visited Russia and France, studying things military. In 1872 he became assistant secretary of war—a position which in Japan is always held by a general officer of the army. In the following year he was made lieutenant general and two years later secretary of war.

The next year saw Japan in the throes of a fierce civil war. The rebellion was led by her greatest soldier, Field Marshal Saigo, who had with him some 50,000 of her best trained samurai. The government was compelled to put forth its greatest strength. An imperial prince was appointed to the nominal command, but as chief of staff Yamagata was the real general in chief and led the forces which crushed the rebellion. Saigo, having been slain, Yamagata became the first military man in the empire and was promoted to the full rank of general.

Being a man of great mental ability, boundless energy and strong personality, he soon became almost as prominent in the political world as in the military and shared with Marquis Ito the position of greatest influence with the Emperor. He was several times prime minister and when not in that position always held some portfolio in the cabinet. He never ceased his active share in the development of the army. Through various official positions, such as inspector general, chief of the general staff and secretary of war, he kept himself in close touch with all parts of the army organization.

When war started with China in 1894 Yamagata was immediately given command of the first army that invaded Manchuria. Those who were with the army at the time describe the immense enthusiasm with which the coming of the great general was greeted by his soldiers. The rigors of a Manchurian winter speedily reduced Yamagata to such a condition that the Emperor, fearful of losing altogether the service of his ablest officer, called him back to Tokio to act as his chief military adviser.

After the war Yamagata was made a marquis and the new military rank of field marshal was established, to which he was promoted. The active interference of Russia, backed by France and Germany, which deprived Japan of the fruits of her victory, led the government to try to come to some understanding that would preserve the independence of Korea. Yamagata was appointed special ambassador for this purpose and proceeded to St. Petersburg, where he effected the treaty which is the basis of Japan's latest demands upon Russia. Not trusting altogether to this, Japan proceeded to double her army and greatly increase her navy. Several officers were promoted to the rank of field marshal in the army and an equal rank in the navy and organized into a supreme council of war.

Of this Yamagata was made chief, a position which he held at the outbreak of the war with Russia.

Field Marshal Oyama, Chief of Staff.

Field Marshal Oyama of Japan is a member of the Supreme Council of War and chief of the general staff of the Japanese army. Oyama has steadily risen in the military organization of Japan with Yamagata, and if the latter has been the Emperor's right-hand man Oyama has been his left. Like Yamagata, he began his career in the war of the restoration. In 1872 he was promoted major general and sent to Europe, where he spent three years studying military science. Returning in 1875, he was made vice minister of war. He added much to his reputation by his skillful conduct of operations in the rebellion of the great Saigo.

In 1879 he was promoted lieutenant general. Two years later he was made minister of war. In 1883 he became chief of the general staff. From this time until the outbreak of the war with China he occupied various positions in the cabinet, usually minister of war. In 1890 he was made full general—a rank then held only by himself and Yamagata. When war started he was looked upon as certainly as Yamagata to command one of the armies. Accordingly he was given the command of the second army of invasion. Being five years younger than Yamagata and more robust, he proved fully equal physically to the task, and conducted the extensive campaign that resulted in the capture of the two great Chinese strongholds—Port Arthur and Wei-hai-Wei. Both positions were considered impregnable.

The mathematical precision with which Oyama conducted the operations and the gallantry with which he took redoubt after redoubt, until finally he swept into the great fortress itself excited the admiration of experts, naval and military. Crossing the Gulf of Pe-Chee-Lee with an army of 30,000 men, he swept the Chinese forces out of Wei-hai-Wei with the greatest apparent ease, and with the aid of the navy cooped up the Chinese fleet within the harbor, where it was quickly destroyed or captured. Oyama, therefore, returned from the war full of honors, was promoted along with Yamagata to the special rank of field marshal. Oyama is noted for his geniality, his calmness under trying circumstances and his studious character.

Lieutenant General Count Nodzu.

After Yamagata and Oyama the most conspicuous military man in Japan is General Nodzu, who succeeded Yamagata in the command of the first army of 1894-95. Like the others, he began his career in the war of the restoration, in which he served as captain. Five years later, at the age of 30, he had reached the rank of colonel. In 1876 he visited the United States, attending the centennial exposition and making a study of the American military system. In connection with the latter he took part in an Indian campaign. He returned to Japan just in time to take part in the civil war inaugurated by Saigo. Promoted major general and given command of the Second brigade, he rendered particularly brilliant service. In 1886, in company with General Oyama, he made an extensive tour of military inspection in Europe and America.

Upon his return he was promoted lieutenant general and placed in command of the Fifth division. When war started with China this was the first division to take the field. A mixed brigade was sent over and beat the Chinese at Asan, in southern Korea. Shortly after this Nodzu, with the remainder of the division, arrived and, finding that the enemy had concentrated a force of some 20,000 men at the famous stronghold of Phyng-Yang, in northern Korea, he moved rapidly against it and crushed it in battle. Another division was now sent over to join him and the two divisions, together about 45,000 strong, became the first army of invasion, of which Yamagata took command.

With this army Yamagata crossed the Yalu river and invaded Manchuria, but his health failing, he was soon forced to return to Japan, whereupon Nodzu was given the command and promoted to the full rank of general. In a very tedious and difficult winter campaign he pushed his way across the southern part of Manchuria, driving the Chinese before him and beating them whenever he could come in touch with them until he reached New-Chwang, where a great battle terminated the campaign.

For his conduct in this war General Nodzu was decorated, made a count, pensioned and honored in various ways. With the enlargement and reorganization of the army the country was divided into three grand military divisions, eastern, middle and western, and General Nodzu was given command of the eastern, which, at the age of 60 years, he still held at the beginning of the war with Russia.

Prime Minister Katsura.

Lieutenant General Katsura, prime minister of Japan, is a man of great ability and energy and a thoroughly trained soldier. He was 20 years old at the time of the restoration and served as a staff officer. He did so well that he was rewarded with a sword of honor and a pension. It was only natural that so promising a young soldier should be one of those chosen to go abroad to study, and in 1870 he was sent to Germany, where he studied for three years. Returning in 1873, he took part, with the rank of major, in the expedition sent to chastise Formosa.

On his return he was designated military attache to the legation in Germany, where he remained for several years. Upon his return he was made lieutenant colonel and appointed director of the army intelligence bureau. Being a man of great capacity for work, he was also made a member of the committee for the investigation of the coast defense works and also given the political post of chief secretary of the cabinet. That he should have held all these offices at once is a high tribute to his ability. In 1882 he was promoted colonel. In 1884 he traveled in Europe on a tour of military inspection, returning the year following. He was now made major general and entered the war department as director of the general affairs bureau.

In 1886 he became vice minister of war. In 1891 he received the rank of lieutenant general and took command of the Third division. In the war with China he was ordered to Korea, where his division, together with the Fifth division, constituted the first army and did brilliant service in Manchuria. Katsura was General Nodzu's right arm in that campaign.

Upon his return he was decorated, made a viscount and given a life pension. Later on he was promoted to the rank of full general. He was then appointed governor general of Formosa, a position of mixed civil and military duties that made it similar to the same office in the Philippines. Some time later he resigned this post to enter the cabinet as secretary of war. Later, at a time of particular political stress, he was invited by the Emperor to form a cabinet, and he succeeded.

At the beginning of the war with Russia General Katsura was 56 years old, ten years younger than Yamagata, and five years younger than Oyama and Nodzu.

Viceroy Alexieff.

Admiral Eugene Alexieff, the Russian Viceroy in the far East, has had a remarkable career, and it may be said to have begun in the United States. When he was only a lieutenant in the navy he served four years on a vessel exploring the coast of northern Siberia. Obtaining a leave of absence he started home by way of the United States. At San Francisco he learned that Russia had abrogated the clause in the treaty of Paris limiting her right in the Black Sea. It was for this that the Crimean war had been fought, and as a result of Russia's action war was again imminent between England and Russia. Alexieff cabled his government for permission to purchase commerce destroyers in the United States to prey upon British commerce and this act put him in high favor with the authorities at St. Petersburg who at once granted his request. He purchased light big steamers and had them ready with steam up to begin their work as soon as war should be declared. The crisis was averted, however, and Alexieff disposed of his vessels to good advantage, spent a part of the season at Long Branch, and upon his return to Russia was given command of a ship and the rank of captain. Three years later he was made an admiral. His close friendship with the Czar was formed when he commanded the vessel upon which the Russian ruler visited the far East, and resulted in Alexieff's appointment as Governor General of Russia's far Eastern provinces. He was in command of the Russian forces which, with the troops of other nations, went to the relief of the embassies at Pekin at the time of the Boxer war. He served for a time as minister of war, and was then appointed viceroy in the East with powers almost as great as the Czar himself. In personal appearance Viceroy Alexieff is six feet tall, broad shouldered and wears a thick brown beard tinged with grey. He was sixty years old at the outbreak of the war with Japan.

Kouropatkin, Leader of the Army.

General Alexei Nicolaievitch Kouropatkin, who was sent to the far East as the commander in chief of the Russian forces early in the war, was one of the most trusted, most powerful and most faithful servants of the Czar. Long ago, when Kouropatkin was a young man, he was sent into Turkestan with the Russian advance and spent his youth in high adventure and in winning for himself honors, decorations and promotion. When he had served his apprenticeship in the marches and bivouacs of the East he returned to Europe and resumed his studies in the schools. At Paris he won the Legion of Honor, and was the first Russian officer to do so. This was in 1874. The following year he went to Kokand, fought in the Pamirs—the "roof of the world"—and in the country of Kublai Khan. He rode 2,500 miles on horseback, came back, wrote a book and won the gold medal of the Geographical Society. He entered the war with Turkey as a lieutenant and emerged from it a colonel. Slowly he rose on the ladder of advancement, until in 1898 he was made absolute master, under the Czar, of the armies of all the Russias.

Kouropatkin was the right-hand man of Skobeleff all through the Russo-Turkish war. Kouropatkin became the hero of the Russian army, second only to his great leader, Skobeleff, by his bravery and fine generalship at the capture of Geok Tepe in 1882.

When the Russians, balked in their dreams of winning Constantinople by the Berlin congress, were making their great swoop through central Asia to the gates of Herat, Lord Salisbury told the British public not to be alarmed for the safety of India. "They will not be able to conquer the Turcomans," he declared. "The Turcoman barrier will last for our lifetime at least." General Tergoukasoff, the Russian commander in central Asia, disagreed with Lord Salisbury. He told the Czar that the Turcomans might be conquered by three years' hard fighting. "That is too long," said the Czar. He recalled Tergoukasoff and sent Skobeleff to command the troops. Skobeleff promptly secured Kouropatkin for his chief lieutenant and together they performed in a few weeks the task which the British premier had declared would take a lifetime.

Archibald Forbes, the war correspondent, was fond of telling how he met Skobeleff, the Russian general, after one of the fiercest of the many desperate fights before Plevna. "I was sitting in my tent writing a dispatch," said Forbes, "when the flap was suddenly drawn aside and in stalked the most terrible and awe-inspiring object I have ever seen in my life. It was Skobeleff, whom I knew well, but I had to look twice before I recognized him. His smart general's uniform was torn into shreds and stained with blood and gunpowder from head to foot. His sword, which he held in his hand, was simply smothered in blood, and great drops of it fell on the floor of the tent as he greeted me. There was a terrible gash across the top of his forehead, and his eyes still blazed with the fierce excitement of the hand-to-hand fight which he had just had with hundreds of Turks.

"While he stood there telling me about the battle, his favorite captain, Kouropatkin, came up and called him away to decide about the disposition of some of the prisoners. Kouropatkin looked even more like a god of war fresh from the scene of carnage. He was bleeding from a dozen wounds, but he stood as steady as a rock when he saluted Skobeleff. The latter suggested that he had better go into the hospital, but he curtly replied: 'No, general. There is work to be done.'

"I heard afterward that Skobeleff and Kouropatkin had fought side by side throughout that bloody day, and had slain the Turks literally by dozens. Their exploits formed the theme of many a story told beside the campfires of both armies throughout the campaign."

After the death of Skobeleff in 1882, Kouropatkin was summoned from central Asia by the Czar and given one high military office after another at St. Petersburg, his special work being to reorganize the Russian army. He was regarded at first as a plain, bluff soldier who would never meddle in politics, and consequently everybody welcomed his advancement. But, to the chagrin of the ministers, he developed into an ardent politician and gained great influence with Alexander III and later with the present Czar.

The Japanese Napoleon.

General Baron T. Kuroki, commander of the victorious Japanese troops in the first great battle on the Yalu, is 56 years old, and was already famous for his successful operations in the late war between China and Japan. Kuroki comes of a race of warriors. For centuries back his ancestors have participated in the civil strife between the damyos, or Japanese barons, and it was through the influence of the house of Kuroki that the feudal system was abolished thirty-six years ago.

Emperor Mutsuhito, while acknowledged as the "heaven-born" and coming of a dynasty which antedates the flood, was in a precarious position while the damyos were having things their own way. He was, in short, only nominally emperor, and the real power in Japan was the Sho-Gun, to whom the damyos were subservient. The house of Kuroki brought a strong influence to bear upon the Sho-Gun and succeeded in restoring the power of the Emperor. For this supreme service Mutsuhito is indebted to the father of the present Baron Kuroki.

Rear Admiral Uriu.

Rear Admiral Sotokichi Uriu, who commanded the fleet at Chemulpo when the Russian warships Variag and Koreitz were sunk, quickly became an idol of the Japanese people, coming next to Rear Admiral Togo, who commanded at Port Arthur, in the admiration of the public. Admiral Uriu was educated at Annapolis, and has many friends in America, particularly among the naval officers, with whom he is a great favorite.

Courtesy of Everybody's Magazine.

ADMIRAL ALEXIEFF.
Russian Viceroy in the Far East.

Courtesy of Everybody's Magazine.

GENERAL KOUROPATKIN.
Russian Minister of War and Commander in Chief of the Land Forces.

RUSSIAN OFFICERS.

Courtesy of Everybody's Magazine.
ADMIRAL AVELAN.
Russian Navy.

Courtesy of Everybody's Magazine.
GENERAL WASMUNDT.
Russian Army.

Courtesy of Everybody's Magazine.
ADMIRAL SKRYDLOFF.
Russian Navy.

Courtesy of Everybody's Magazine.
COUNT LAMSDORF.
Russian Minister of Foreign Affairs.

CHAPTER X

ON THE EVE OF THE WAR

How the People of the Japanese Capital Remained Calm in the Face of a Great Crisis, While the Government Secretly Prepared for War—People Knew Nothing of the Merits of the Controversy—Japanese Spies Disguised as Chinamen—Keep Government Informed Concerning Russian Affairs in Manchuria.

THE sights and scenes in Tokio, the Japanese capital, immediately before the war were in great contrast to what might have been witnessed in the capital of any other nation in the face of a great crisis.

Instead of the excitement which marks warlike preparations elsewhere the Japanese people were tranquil and silent. This was due in part to the secrecy which marked all the movements of the government. In fact, so secretly has the government acted that it is doubtful if half a dozen men in all Japan know exactly what all the row was about or could detail the course of negotiations with Russia since the situation became dangerous to peace. There had been notes and notes and replies and replies. Some were principal notes and some were principal replies, and out of it all, if one had the time and patience, a mosaic could be patched up which probably would fairly represent the truth.

No statement of any sort as to the causes of differences with Russia or their character had been made by anyone in authority. The nation was on the verge of war because of the demands upon Russia, the nature of which it does not pretend to know, nor did it care much. The fundamental causes of general hatred of Russia are well known to every Japanese. The authorities had no need to work up a public sentiment for the war. On the contrary, their task was more in the way of repressing the belligerent feeling of the people.

The extraordinary patriotism of the Japanese leads them to support any war, whatever the cause.

Watching Everything and Everybody.

There was a carefully directed campaign of scrutiny, which involved watching everything and everybody. Private and press telegrams were scrutinized closely for some time, and a secret censorship was in force. The man who filed a press telegram had no means of finding out whether his message had been sent or not, and if it were sent, he had no means of ascertaining whether it had been garbled by the censor. No one knew who the censor was or the location of his office, and it was useless to try to find out anything about it from the telegraph people.

Private messages which were prepaid were in a slightly better case. When words were erased from them the sender was notified in the course of a few days, and the money for those words refunded. He at least had the satisfaction of knowing that his message had not gone as he wrote it.

Preparing for War.

But all the time war preparations went on rapidly. There is no question that the Japanese authorities knew exactly what they had to do to beat Russia. Their military information was wonderful. When they went to war with China ten years ago they not only knew the character of their enemy, but they knew the topography of the country over which they expected to fight and knew the condition and equipment of the Chinese troops.

They were prepared against the Russians fully as well as against the Chinese, if not better. They knew as well as the Russians how many Muscovite troops were in Manchuria and Eastern Siberia and where they were stationed.

Ever since the Russianization of Manchuria began the Japanese have looked forward to the time when they would fight, and have prepared for it. They have studied the country in minute detail. Their maps show the results of this work. Individual huts and clumps of trees were shown. Their knowledge of the country was complete. It is doubtful if the Russians were as well informed as to the topography of the battle ground, both in Manchuria or Korea, as the Japanese.

Jap Spies Disguised as Chinamen.

For some years there have been in the neighborhood of 10,000 Japanese in Manchuria and 30,000 in Korea. Nearly every one of them has been a source of information to the military authorities here, and not a few of them have been military men in one disguise or another. With false pigtails and in Chinese dress they have worked as servants for Russians, understanding and noting every word their masters said.

It is an advantage the Russians can never have. They are forced to rely upon Chinese for spies or upon the very few renegade Japanese they have been able to find, with the added difficulty that such Japanese are almost as well known to their own government as to the Russians.

In the preparation of their war maps the Japanese have a way of making them on a rather small scale, showing a large field of operations; then smaller maps show more in detail sections of the larger. Still smaller sections are shown in still greater detail.

Such estimates as the Japanese war authorities permitted to become known fixed the number of Russian troops available as approximately 200,000, including the railway guards.

To meet this force the Japanese relied on their regular army at the beginning.

The Japanese Army Organization.

The army is organized on the skeleton plan, each company in peace times numbering about half what its full war quota is. There are twelve regular divisions and the imperial guard, which constitutes a division by itself. Each division consists of two brigades of infantry, each brigade being composed of two regiments. The regiment is organized like our own, of three battalions of four companies. But the companies are nearly twice the size of ours when in active service, numbering on the war footing 240 men, so that a battalion is about 1,000 strong.

Besides the infantry, each division has a regiment of field or mountain artillery, two battalions of three batteries, six guns in a battery, 120 men to a battery. The guns are of Japanese invention and make and are capable of doing fine work.

Each division also has a regiment of cavalry, such as it. The horses of the Japanese are very poor, and the men are not specially skillful as riders. Military observers rate the cavalry as very poor. Certainly it is not nearly the equal of the Russian Cossack force. Each regiment consists of three squadrons—troops we should call them—of but 150 men.

There is also in each division a battalion of engineers, who are among the best soldiers of Japan, very highly trained. Besides these there are regular commissariat and supply trains and the sanitary or medical corps.

All told on a war footing each division consists of about 15,000 men. This made the fighting line number about 200,000.

In peace the army is hardly half that size. Every man on reaching the age of 21 is required to serve with the colors, but there are many exemptions in peace times. After three years with the colors the men go into the first reserve for five years. In war the fighting line is at once filled up to the limit from the first reserve.

The Japanese had the inestimable advantage of a well-protected interior line of communication with Korea. From Muji to Fusan it is only a night's steam. Midway lay Tsushima, with its great fortifications. Flanked on either side by the Japanese ships the Russians would have had to exert their entire naval strength to force the passage to get at the Japanese transports.

Constant Naval Practice.

The Japanese refused to permit newspaper men to accompany the first operations, either naval or land, and issued very stringent regulations covering the case. Commanders of naval stations or of fleets were empowered to establish what are termed strategical sea areas into which no ship could come without permission. The captain who endeavored to enter such an area against the wish of the commander was sent back in charge of an armed vessel and imprisoned and fined.

There are three great units in the Japanese navy—the battle ship squadron, the armored cruiser squadron and the cruiser squadron. For months each squadron had been maneuvering by itself, with frequent grand maneuvers embracing the entire navy. At the naval station at Takeshiki, in Tsushima, twenty torpedo vessels had been practicing in one flotilla. The result was that every officer of every vessel knew not only what his own ship would do under given circumstances, but also what every other ship would do.

That intricate and valuable bit of naval information, the helm angle of each ship, was a matter of common knowledge. They maneuvered as well in the dark as in the light, and if a ship was transferred from one squadron to another she would come back to old mates who knew her well, and no new drill was necessary to fit her to the new conditions. Moreover, all the ships were in fighting condition.

On paper Russia was nearly, if not quite, as strong as Japan. In battleships she was one if not two units stronger. In armored cruisers she did not compare. In cruisers she was approximately as strong, but in torpedo boats the Japanese excelled.

CHAPTER XI

THE FIRST SHOT OF THE WAR

A Russian Cruiser and Torpedo Gunboat Trapped in the Korean Harbor but Forced to Fight in the Open—Japan's Second Naval Victory, in which not a Japanese Life was Lost—Recalls Admiral Cervera's Brave Dash at Santiago—First Shot in the Preliminary Skirmish Fired to the Russians.

FOLLOWING fast upon the news of the destruction of the three great Russian warships in the roadstead of Port Arthur came reports of another Japanese victory outside the harbor of Chemulpo, the principal port on the west coast of Korea.

A squadron of Japanese warships was escorting transports loaded with troops from Nagasaki to the Korean harbor when they encountered at the very mouth of the harbor the Russian cruiser Variag and the torpedo gunboat Korietz. Both ships took refuge in the harbor under the fire of the Japanese squadron, but not until the Korietz had launched two torpedo tubes which were ineffective. The Japanese immediately opened a heavy fire and the Variag came to the rescue of the Korietz. Before any serious damage was done the Russian ships retired precipitately into the harbor.

This incident took place on the afternoon of February 8 and marked the first opening of hostilities. The first shot in the war was fired outside the harbor of Chemulpo, although the battle in which the Russian cruisers were sunk did not take place until the next day.

Early the following morning, Tuesday, February 9, the two Russian sea fighters, which had repaired their damages during the night, made a dash out of the harbor. It was a desperate effort to escape from the watchful Japanese fleet, resembling in its hopelessness and dash Cervera's memorable rush from the harbor of Santiago.

The guns of the Japanese squadron covered the entrance to the harbor where the Russians had taken refuge. As the Czar's battleships emerged, belching shot and shell, the Japanese opened fire on them.

The battle was as brief as it was furious. The Japanese concentrated a terrific fire on the two Russian ships and in a very short time it was apparent that their destruction was a certainty.

Shells disabled the Variag, her steering gear was knocked out of commission, her gun turrets battered and within half an hour after the morning's engagement began she sank.

The Korietz fought until a shell exploding in her magazine rent her asunder. The crew of the two ships struggled into the water and the survivors were picked up by the French cruiser Pascal, which witnessed the terrific battle.

Many officers and marines from the Korietz and the Variag made their escape to the Korean shore, where they were captured by the Japanese patrol established there.

The Japanese squadron proceeded on its way to Chemulpo, where the troops on board the transports were landed.

Story of an Eyewitness.

One of the eyewitnesses to the battle was a London war correspondent, whose account is exceedingly graphic. He says that the Japanese landed 2,500 men on the afternoon of February 8 and on the next morning the Korietz and Variag were ordered to leave port before noon. The correspondent said: "At 11:30 they steamed away. I proceeded in the steamer Ajax, from which I saw them met by eight Japanese vessels. The first gun was fired at 11:40 a. m. The Japanese, scorning the Korietz, which was untouched, concentrated their fire on the Variag. The latter continually circled round, replying from her sides alternately, but it was apparent her shooting was not good. On every side her shells went wide. It is observable that the Japanese gradually closed, the battle ship Mikado doing the most firing and effecting the most damage.

"At 1:15 they ceased firing. As the Variag re-entered the harbor she took up a position with the Korietz among the other warships. One of her boilers was injured and she was on fire astern. The flames were extinguished by flooding a compartment. She refused to disclose the number of her casualties.

"The Japanese, meanwhile, had withdrawn again, waiting until 4 o'clock.

"My writing is interrupted at this moment by a terrific report. The Korietz has been blown up by the Russians, whose men can be seen in boats pulling for the Variag. An immense column of smoke arose and then cleared away, giving a sight of the Korietz with funnel masts just above the water.

"The Japanese ashore are wildly cheering. The Variag still remains at anchor.

"It was exactly at 4 o'clock that the Korietz exploded. Within half an hour the Japanese fleet appeared in the dim distance and approached slowly.

"At 5:20 p. m. fire appeared in the afterpart of the Variag and spread slowly.

"The Japanese then stopped firing. The Variag heeled over, surely but barely perceptibly, and at 6:05 p. m. she sank with a dull rumble.

"All the men of the Variag and Korietz were removed by the French cruiser Pascal.

"The Russians now admit that the Korietz fired the first shot, but say it was accidental. No doubt they made a fine fight against odds.

"The conduct of the Japanese everywhere was exemplary. They express sympathy with their foes, but say they are compelled to take extreme measures."

Official Report of the Battle.

The text of the official report of the Chemulpo affair is as follows: "On Monday a Japanese squadron escorting transports met on the way to Chemulpo, Korea, the Russian gunboat Korietz as the latter was coming out of port. The Korietz took up an offensive attitude toward the Japanese vessels and fired on the Japanese torpedo boats. The latter discharged two torpedoes ineffectively and then the Korietz returned to her anchorage in the port.

"Early in the morning of Tuesday Admiral Urik, commanding the Japanese squadron, formally called on the Russian warships to leave Chemulpo before noon. The admiral added that if his demand was not complied with he would be compelled to attack them in the harbor. The two Russian warships left the port at about 11:30 a. m. and a battle followed outside the Polynesian islands. After about an hour's engagement the Russian warships sought refuge among the islands. Toward the evening the Russian cruiser Variag sank and at about 4 a. m. to-day, February 10, the Korietz was reported to have also sunk, having been blown up. The officers and men of the two sunken vessels sought refuge on the French cruiser Pascal. There were no casualties on the Japanese side."

The cruiser Variag, which was built at Cramp's shipyard, Philadelphia, Pa., in 1893, was of steel and unsheathed. She was of 6,500 tons displacement and her indicated horse power was 20,000. On her speed trials she made twenty-four knots per hour for eight hours. She was 400 feet long, 52 feet in beam, and had a depth of hold of 20 feet.

The Variag's armament consisted of twelve 6-inch quick-firing guns, twelve 12-pounder quick-firers, eight 3-pounder quick-firers, two 1-pounders, and six torpedo tubes, two of which were submerged.

The Korietz was built in Stockholm. She was of steel and was 206 feet in length, 35 feet in beam, 1,413 tons displacement, and 1,500 indicated horse power. Her speed was thirteen knots.

The armament of the Korietz consisted of two 8-inch breech-loaders, one 6-inch breech-loader, four 4.7-inch quick-firers, two 6-pounder quick-firers, four 1-pound revolving cannon and two torpedo tubes.

The Variag sank as the result of the damage inflicted by the Japanese guns. The Korietz was blown up by its own crew in the harbor in order to prevent capture by the Mikado's fleet.

Bishop Moore's Graphic Description.

The most graphic and thrilling account of the naval battle at Chemulpo was furnished by Bishop David H. Moore, of the Methodist-Episcopal Church, whose missionary field is in China, Japan and Korea. The author is personally acquainted with Bishop Moore, and knows him to be a man whose powers of observation and eloquent discription probably exceeded those of any other eye-witness of this engagement.

The Bishop left Shanghai February 6 in the Sungari of the Russian line, plying between that city and Port Arthur, which cast anchor in Chemulpo harbor on the morning of Feburary 8, near the Russian cruisers Variag and Korietz. Not a Japanese cruiser was in sight at this time.

Bishop Moore's account of the arrival of the Japanese fleet, and the subsequent action, is as follows:

"Nothing unusual occurred until about 4 p. m., when a long line of Japanese cruisers, torpedo boats, and three transports—twelve in all—steamed in and anchored. It was dark before the troops began to disembark—some three thousand hardy-looking fellows, equipped for active service. The soldiers landed, the fleet withdrew some eight miles beyond an island, and formed across the narrow channel by which alone heavy-draught ships can pass in or out—the Asama, Takashito, Gushirna, Nanivo, Chiyoda, Nitoka and eight torpedo boats. Besides these, a second line was rumored.

"Tuesday, the 9th, like a shock ran through the city the report that the Japanese consul had notified the Russians that if they did not sail out by noon, they would be attacked in the harbor at 4 p. m. The Russians decided to go out at once and stripped their ships for action.

"Our mission compound commands a fine view of the harbor and roadstead, and to the left and further seaward a still better view is secured. As the ships disappeared in the haze, our hearts stood still with almost agonizing suspense. Then came the roar of two shots across their bows.

"Then refusing to 'lay to' they opened their batteries and were opened upon by the concentrated fire of the Japanese. Fifteen minutes we thought would suffice to end the unequal combat, but earth and sea shook under the awful thunder of the guns, thirty minutes, forty-five, fifty-two minutes, and unable to break through, scorning to surrender, the Russians swung around and steamed back to their anchorage, with their flags still flying. Sure of their prey, and perhaps unwilling to fight unnecessarily in the harbor, the Japanese did not grumble, but resumed their station in the roadstead, completely blocking the only channel. The four-funneled, Cramp-built big cruiser, Variag, was evidently badly damaged and listed to port. The Korietz, the smaller of the two, was apparently uninjured.

"We hastened to row out in a sampan to inspect for ourselves. We saw no scars on the Korietz, though the sailors were putting fresh paint on her hull here and there, as if to conceal scars and the officer directing had his head bandaged. Allen (United States Minister) asked in Russian how they fared. The commander replied that they had no chance, and that at 4 p. m. the ship would go up. The Variag was evidently sinking. She was mortally wounded amidship and had a huge rent in her upper works. Two of her funnels were riddled, and her bridge was a mass of twisted iron.

"A lieutenant who was on the bridge when it was struck, was torn to pieces and blown overboard, all but his right arm and hand, which were found still holding the flag with which he was signaling orders to the Korietz. The boats of the other warships were removing her men to the security of their own decks. The wounded were being taken to the French and English ships—a hundred more dead than alive. It was pathetic, the tenderness and veneration with which they handed down the Czar's portrait. Our United States ship, the Vicksburg, alone gave no sanctuary, though her lifeboat helped remove the men.

"We hailed the first officer of the Sunbari, the ship we landed from the day before. He indicated that all was lost, and shortly after we saw the men dash below as if to scuttle her.

"Now, the men are hastening to leave the Korietz. We are within a few yards of her last two boats as they put off. It is twenty minutes till 4, and we recall her captain's words and hasten our rowers. There is an island surmounted by a revolving light, 600 yards away. We land and climb to its summit. The hands of the watch denote 4. Instantly a terrific explosion in the stern, and almost simultaneously another forward, sent the Korietz to its doom. Two malignant volumes of smoke and debris leap, writhing and twisting upward, clinching and struggling, as though two monsters in mortal combat. And as their black bodies pulled apart for a moment, the sinking sun, tearful with filmy haze, shone through.

"And listen! Through the blackness of darkness and the rain of falling fragments of their ruined ship come cadences stately and solemn and grand from the French ship, where they had asylum, her crews joining in the majestic chant of the Russian national hymn, at once their new oath of allegiance to the Czar and a requiem over their lost ship. The smokestack, her gleaming prow, and portions of her steel frame, show where the Korietz met her fate.

"Now a fierce fire rages in the bunkers of the Variag, more and more she lists to port. She has outlived the sun, but at 6 o'clock, with one great shudder, like a huge leviathan, she turns on her side and dies.

"Only the Sungari remains, so recently our home. She sinks, all too slowly. A boat puts off to her from the French cruiser, and soon her beautiful upper works are a roaring furnace of flames. All night she burns and glows, and dies with the morning light.

"All this in 'a state of hostility.' What shall we see when war is formally declared?"

Bishop Moore's thrilling description thoroughly corroborates all reports of the heroism displayed by the crews of both the Russian ships, which greatly impressed all the foreigners present, who believed the Russians were going to certain destruction. As the Variag and the Korietz steamed past the foreign war vessels the crews manned the sides and cheered continuously and the strains of the Russian national anthem accompanied the Russian sailors as they went into battle.

Bravery of Russian Sailors.

The Russian officers and crew behaved splendidly. Three times flames broke out on board the Variag while she was under fire and each time they were extinguished as coolly as if the men had been at drill. The wounded men were carried below and the members of the crew who lost their lives were replaced by others. The holes made in the Variag by the projectiles of the enemy were not stopped up. The Variag sustained terrible damage. Both starboard and port bulwarks were destroyed, her guns were crippled and her hull was riddled with shot. Thirty-three men on board of her were killed in the first attack, including Midshipman Count Nirod.

The Japanese squadron lost a number of men. Two Japanese cruisers were damaged and one torpedo-boat was sunk.

The official report of the Chemulpo fight gives the losses on the Variag at one officer and thirty-three men killed, two officers slightly wounded and seventeen men severely wounded. There were no losses on board the Korietz.

The charge that the American cruiser Vicksburg had refused to aid the wounded Russians was disproved by Captain Marshall's report to the Navy department, in which he said that he was the first to send medical assistance to the Variag as soon as he learned that the Russians were abandoning the vessel. He also sent three boats and assisted in taking off the Russian sailors and putting them on board the British and Italian vessels. He also offered the use of the American transport Zafiro, which was declined.

CHAPTER XII

PORT ARTHUR, THE RUSSIAN "GIBRALTAR"

Named by the English for a Daring Naval Lieutenant, Fortified by the French for the Chinese, Won by the Japanese in the War With China and Finally Leased to Russia—Dalny, the Commercial City, Splendidly Located, to Be Terminus of the Siberian Railway—Port Arthur as a Purely Military and Naval Base.

THE famous harbor with a neck like a bottle, whose waters wash the shores of the southern extremity of the Liaotung peninsula, where Japan struck her first serious blow against Russia—which with the great fortress on its cliffs is known as Port Arthur, has had a varied and most remarkable history.

For hundreds of years Chinese-coasting junks beating along the Yellow Sea in the coastwise trade had run into the land-locked harbor of Lu Shun Kow, down at the extreme southern end of the Liaotung peninsula. All along the shore great grey cliffs ran up straight from the sea to a height which varied from 300 to 1,500 feet. If you came close enough in you could make out a slit in the mountains which gave entrance to a body of water within. This slit was not more than 200 or 300 yards wide. Once you got through the passageway there was a wide stretch of water before you, hemmed in by cliffs on every side. At high tide the water was deep enough for the anchorage of a big ship, but when the tide went out hundreds of acres of mud flats were exposed to view.

In the valleys between the cliffs were built some fifty or sixty miserable mud huts, and in them lived 300 or 400 Chinese coolies. That was the situation up to 1868, in what is now Port Arthur.

Daring Feat of Lieutenant Arthur.

In that year three or four ships of the British navy came that way on a surveying expedition. One of these ships was the gunboat Algerine, commanded by Lieutenant William Arthur. Lieutenant Arthur daringly ran his vessel in between the cliffs which guard the harbor, which was thereupon named Port Arthur, in honor of his exploit.

But beyond the name, Port Arthur gained no new fame for another twenty years. In 1821 it was merely a convenient harbor into which coasting junks could run for safety when great storms swept the sea outside. On the cliffs and in the valleys thereabouts there still lived only a few hundred wretched Chinese coolies.

Then the great Celestial empire began to wake up. Foreign engineers were sent along the coast to pick out a safe harbor which might be fortified and made the chief station for the new and modern navy of China. They settled on Port Arthur, and it was planned to transform the place into an immensely strong and completely fitted naval station. Plans were drawn for great dockyards, workshops, dry docks, refitting basins and foundries, while above them on the commanding cliffs strong fortresses were to be erected. The contracts for all this work were let to French contractors, so that it was France which first among the nations had to do with this Gibraltar of the far east. French contractors, with the aid of swarms of Chinese coolies, working like slaves for 15 cents or 20 cents a day, worked for years on the works in and about Port Arthur. Not until 1891 was the place turned over to China ready for occupancy as a great naval station.

By that time Port Arthur—its namesake already forgotten—had become a fairly well built town, containing more than 1,000 houses and shops, outside of the government works. It then had a commercial population of 6,000, to say nothing of the Chinese garrison of 7,000 or more soldiers. The forts were mounted with modern guns, and Chinese gunners were trained by German and other foreign experts in the use of the artillery.

Captured by the Japanese.

In 1894, during the month of November, the victorious Japanese army marched through one of the narrow passes which lead through the surrounding mountains to the City of Port Arthur, captured the city and put to the sword many of the inhabitants, non-combatants as well as members of the garrison. It was a bloody day, though Japanese officers stopped the slaughter as quickly as possible.

For a time, then, Port Arthur was apparently in the permanent possession of the Japanese, until the pressure of the allied powers forced her to give it back to the Chinese in January, 1896. Before they marched out the Japanese destroyed a large part of the Chinese fortifications.

In 1898 Port Arthur was "leased" to Russia, which immediately fortified it, with the intention of making it the strongest port in the eastern seas. Its importance to Russia is great. Vladivostok, the other great Russian port on the Japan Sea, is icebound a large part of the year. The possession of Port Arthur gives the navy of the Czar a port which is never frozen. Moreover, it is a port which commands the approach to Pekin, the Chinese capital.

Never since the Russian occupancy has there been any cessation of activity in and about Port Arthur. In miserable hovels on the hillsides swarm thousands of coolies, who at a word can be hired for 20 cents a day to do any kind of hard and adventuresome work. Last year a Russian contractor at Port Arthur offered to bet that within a half hour he could hire 10,000 men outside of his regular large force. These regular forces are extremely large, and are kept steadily at work both by land and sea. Any time within the last five or six years one could find in the outer harbor a fleet of 500 to 1,000 Chinese junks, all loaded with railroad ties, lumber and other building material. They, of course, are all working for the Russian government.

Large Purchases of American Goods.

One important result of the Russian occupancy of Port Arthur has been a tremendous increase in the imports from the United States. During several weeks in 1902 American goods to the value of more than $2,000,000 weekly were landed at Port Arthur, and

Courtesy of Everybody's Magazine.
VICE-ADMIRAL ENOMOTO.
Japanese Navy.

Courtesy of Everybody's Magazine.
FIELD MARSHAL OYAMA.
The "Left Hand of the Mikado."

Courtesy of Everybody's Magazine.
GENERAL NOZU.
A Japanese Hero in the War With China.

BARON TADASU HAYASHI.
Japanese Minister to Great Britain.

the yearly commerce of the United States with that and the adjacent ports has been estimated at nearly $100,00,000.

But the Russian plan had been from the first to make Port Arthur a purely military and naval center. With that plan in view, the Russians several years ago began the construction of the wonderful City of Dalny, thirty miles north and ten miles east of Port Arthur, which they hoped to make the commercial capital of the far east. The plan contemplated that all commercial ships should be barred out of Port Arthur and sent to Dalny, and that the former fortress should be barred to civilians, where, indeed, they have been allowed only on sufferance, properly being held on merely temporary leases.

Dalny—or rather the site of the present city—was located on an open roadstead, where the navies of all nations might ride. In order to make there a safe harbor, an immense breakwater, costing millions, was built and is now completed, projecting into the sea for a great distance and enclosing a splendid anchorage. At Dalny, also, great administration buildings were erected, and even —that rare thing in the far east—a first-class and comfortable hotel.

Eventually, as planned, Dalny is to be the final terminus of the great Siberian railroad, by means of which Russia has tied together her widely extending empire.

Visitors to Port Arthur within the last few years have been vastly impressed by the spirit of boundless energy which prevails there. Life in the fortress city is in great contrast to that in most of the settlements along the Chinese coast. The streets have been thronged with Russian soldiers and with gangs of coolies, all busy on some important errand.

The Little Cross of St. George.

The Russian soldier, as seen at Port Arthur, impresses the visitor as being in deadly earnest. Before them all, from the lowest private in the ranks to the highest officer, shines the hope of winning the little cross of St. George for valor in the face of the enemy. And on the day of St. George the brave men who wear his cross have the honor of breaking bread with the great white Czar himself, in his palace at St. Petersburg, if they be stationed there, or, if they are quartered at Port Arthur, they eat breakfast at the table of the Czar's viceroy, Admiral Alexieff—and how can greater honor come into the life of one of these wiry Cossacks, wrapped in skins and furs and mounted on a shaggy pony, even tougher and hardier than his master?

So, strangely, in the passing of the years and in the working out of the policy of the nations, has the little Chinese junk harbor of forty years ago, named by the English, fortified by the French for the Chinese, won by the Japanese, at a great cost of blood, and finally leased and turned into a Pacific Gibraltar by the Russians, come to be the center of the world's interst.

Russia's Improvements at Port Arthur.

When the Russians entered Port Arthur they found it contained fortifications, dock yards and an arsenal, all erected by the Chinese and built under the direction of foreign engineers. It was considered impregnable even then, yet the Japanese took it from the Chinese by force. It has been admitted that the Chinese fortifications were defective, and it was believed the Russians had made it a Gibraltar, indeed.

As a glance at the map will show, Port Arthur is situated at the southeastern point of the Liao-Tung peninsula, which divides the Gulf of Pe-Chee-Lee from the Yellow sea. Directly opposite, and facing it, 110 miles distant, is Wei-hai-Wei, now controlled by the English. It was believed that a strong naval force, operating from Port Arthur as a base, could control the entrance to the Gulf of Pe-Chee-Lee absolutely and thus become master of the marine highway to Taku, Tien-Tsin and Pekin. This consideration certainly added immensely to the strategic importance of the position.

If anything were lacking in the topography of Port Arthur and its environs to render it an admirable military stronghold the deficiency would be more than supplied by the character of its water approaches. The town is situated on a bay connected by a narrow strip of land with the peninsula proper.

From seaward the port is reached by a winding channel, not more than 300 yards across in its widest part, and narrowing to less than 200 yards in some portions. This channel runs northward from the open sea for three-quarters of a mile, and for nearly the entire distance is enfiladed by a fort carrying a heavy battery, which is located on a curving point on the western shore. This fort was erected originally by the Chinese, and by them was named the "Tiger's Tail." Since it passed into the hands of the Russians it has been strengthened enormously, and another fort on the opposite bank has been reconstructed and fortified so as to command the passage. In the face of the destructive fire that could be poured from these forts, it would be impossible for any naval vessel attempting to enter the channel to live for more than a few moments, even if she escaped annihilation by the mines and torpedoes at the entrance.

Strong Batteries Erected.

Russia, after her acquisition of the place, erected strong shore batteries commanding the bay itself and the approaches to the channel. These batteries consist of heavy Krupps and smaller rapid-fire guns, which were supposed to effectively command the entire entrance of the harbor from point to point, the distance being not more than 1,200 yards.

Inside the heavily fortified entrance to the harbor the channel widens into a broad, shallow stretch of water, which forms the basin on which the dock yards are located. These yards were elaborate as completed by the Chinese in 1890, but, of course, they have been improved and extended since that time. They include a dry dock 400 feet long, work shops and storehouses completely equipped for the building or repair of vessels; foundries and magazines for turning out guns and ammunition and even a torpedo factory.

These various structures extend about three sides of the big tidal basin, and the different departments of the yards are connected by a railway half a mile in length.

A Strong Naval Base.

Port Arthur thus possesses all the natural advantages which go to make a strong naval base. That its fitness for the purpose was appreciated by the Chinese themselves is shown by their selection of it as their chief naval station before the outbreak of the Chino-Japanese war.

At the time the defense of the place was provided for from the sea front by a number of powerful batteries and strongly defended forts. The only stone fortifications were re-enforced by earthworks and some forty Krupp guns, varying in caliber from six to nine and one-half inches, were placed in position to command the harbor. A large number of rifled mortars, howitzers and rapid-fire guns were located on the defenses, which extended over three and one-half miles of the coast line. At the same time the entrance to the harbor was provided with an elaborate mine field, likely to prove destructive to any fleet that tried to force an entrance.

Costly Work of French Engineers.

Port Arthur as a fortified naval base dates back to the last Franco-Chinese war. It was after that war that China resolved to make provision against similar disasters in the future, and in pursuance of this design she determined to transform Port Arthur into a sort of far eastern Gibraltar. The dock and arsenal were built for China by a French syndicate at an enormous expense and not without several mishaps—part of the dock collapsing the first time the water was let in. Port Arthur to-day is not a single fortress, but at least a dozen batteries are scattered over several miles of the coast, and on both sides of the little C-shaped basin.

The Russians strengthened the series of fortifications known as the Hwang-Chinshan forts, which command the entrance of the harbor to the east, and directly behind they rebuilt a chain of batteries which are intended to pour down shot and shell into the inner harbor. The Laomuchu battery is so placed that it sweeps the approach of the port diagonally and commands both the outer and inner basins. The village of Port Arthur is situated opposite the entrance to the harbor.

The peninsula is joined to the mainland by a narrow strip of land, which is only eighteen miles wide in some places.

In 1894 the Japanese landed 40,000 soldiers on the peninsula immediately after the naval victory off the mouth of the Yalu river, cut off Russia's land communications and captured Port Arthur.

Experience of recent wars leads one to question the ability of fortifications to protect a fleet. At Manila, it will be remembered, Dewey's ships were not damaged to any great extent, while they practically destroyed the Spanish ships covered by the forts. At Santiago the American fleet cruised out of range of the forts and threw shells into the harbor. Port Arthur itself, as has been told, fell a prey to an opposing force when the Chinese defended it in 1894, and, as a matter of fact, the Japanesè took it in a single morning.

Although the Russians have blasted and dredged the channel, the harbor, which lies behind a hill, cannot safely be entered by the heavier draught vessels. These are compelled to anchor in the roadstead. In daylight the forts might have been able to reply with more execution than they did during the night attack of the Japanese torpedo-boats.

The big Krupp guns mounted in a fort on the hill to the east of the harbor have great range, and the entrance to the harbor is so narrow that no hostile fleet would attempt to enter it, as none attempted to get into the harbor of Santiago. Yet the distance from the river basin is not so great that a hostile fleet could not stand outside and seriously interfere with it.

Description of the Harbor.

The harbor is an oval inlet of the sea, two miles long from east to west, and a mile from north to south; it is surrounded by hills of varying elevation, and its sole entrance is on the southern side by a narrow channel guarded at the southwestern end by a couple of dangerous reefs and protected against bad weather by a narrow split of rock land known as Tiger's Tail, which runs diagonally across its northern extremity. This harbor, however, was so shallow that until extensive dredging operations had been undertaken no vessel of any size could enter; even now there are berths for but three battleships in addition to smaller craft. For this reason the major portion of the Russian fleet has always been forced to lie outside the heads, or else enter the large swing basin or wet dock, which lies to the east, facing the entrance to the harbor proper.

The approach to the harbor and basin is very confined and from the nature of its surroundings is very easily defended. To the east, Kwang-Chin, or Golden Hill, rises to a height of nearly 2550 feet above the sea level, and its elevation has been taken full advantage of by the erection on and around its summit of three powerful batteries, mounting besides smaller guns, four new breech-loading cannon, weighing sixty-three and one-half tons, on fortress mountings. On the side facing the entrance, and half way down the slope, are two batteries of quick-firers, for the most part Canet 5.5 inch and 75 mm. guns, in addition to a torpedo and searchlight station. The fortifications extend from the Kwang-Chin-Shan fort for a distance of nearly nine miles in the northern direction, and this line is joined by a circle of batteries on the hill-tops surrounding the town to a second long line of defenses, starting south around the peninsula from the Mantow hill. So much for the eastern side of the entrance. On the west the most important fort is the Wei-Yuen, and this is joined to several small quick-fire batteries commanding the entrance by castellated bridges. A short time ago the whole of these fortifications were surrounded by a high wooden palisade to prevent the inquisitive from learning too much.

From Pinnacle Rock.

The width of the entrance from Pinnacle rock, on the west, to the opposite shore is barely 350 yards, while the three-fathom channel at its narrowest is not more than 500 feet in width. Within the heads it widens out somewhat, and between the end of the Tiger's Tail and the entrance to the basin there is a width of 430 yards; even this makes it a most difficult task for any vessel over 300 feet in length to enter or leave either the harbor or basin.

On the Tiger's Tail are placed seven Canet 5.5-inch quick-firers in an open battery at an elevation of not more than ten feet above the sea; at the extreme end of the spit is a quadruple launching slip for destroyers, from which two lots of four have recently been launched after being sent out in sections. Behind this, again, is a circular observation tower and flagstaff.

The basin or east port was excavated primarily by the Chinese, as also the dry dock cut in its northern side. It has an average depth of three and one-half fathoms and can accommodate nearly a dozen large vessels. The western end is devoted exclusively to torpedo craft, though a dock for these small boats is in process of construction on the eastern side. The dry dock, repaired and enlarged by the Russians, is 452 feet over all, 370 feet over blocks, 90 feet wide at the entrance and has a depth on the sill at high water, ordinary spring tides, of 32 feet. These figures are interesting, for they show that, even with her draught augmented nearly six feet, the Retvizan might have entered the dock for repairs at high water. An eighty-ton sheerlegs is in position on the land side of the basin, and immense engine shops and repairing houses have been constructed wherever there was an available plot of land. Just inside the Tiger's Tail the mud has been dredged away so as to allow destroyers to lie right alongside the building slip. The artillerymen—the garrison gunners in the big shore batteries that frown from every hill—can shoot well, and many of the cannon are of great size.

The Claw of the Great Bear.

Port Arthur, at the tip of the Manchurian peninsula, between the bay of Korea, and the Gulf of Pe-Chee-Lee, is often likened to a claw of the great bear threatening the heart of the Chinese empire. It is within easy striking distance of Pekin and equally favorable for attack on the Korean capital. While other nations have established themselves here and there along the coasts, Russia has lunged the body of its empire into these territories.

Here hostilities began with the brilliant torpedo coup of the Japs, destroying the flower of the Czar's fleet.

Before it was taken by the Russians a few years ago, Port Arthur was a naval arsenal of the Chinese, under the name of Lu-Shun-Kou. The waters of the gulf, entering between two high hills, expand into a harbor which is excellent, though of limited capacity. It is said that not more than four large battleships can find room to maneuver there. Many improvements have been made and more attempted. The entrance has been deepened and the harbor dredged to thirty feet. There are several costly dry-docks, quays and a graving dock for torpedo-boats.

On account of the great mud flats exposed at low tide there is much typhus, and it has been proposed to remedy this, as well as enlarge the roadstead, by making another entrance through the southwest arm of the port. This would afford a circulating current from the sea. A reef of rocks protects the occupants of the harbor from wind and hostile attack. However, this protection, especially from the guns of an enterprising enemy, has been debated, and it seems likely that from the land side Port Arthur could be cut off. At the time of the writer's residence in China naval experts considered Wei-hai-Wei better situated. The first modern improvements at Port Arthur were carried out for the Chinese by a French syndicate more than a dozen years ago.

Forts on Every Hill.

Frowning forts occupy every hill and the soldiers swarm everywhere. At last accounts, and the Russians do not favor knowledge of such things, forbidding visitors entry to the forts, the batteries consisted in part of thirty-five twelve-inch guns, forty-four six-inch and fifty-two four-inch rapid-firing guns. The twelve-inch weapons have a range of more than seven miles. The barracks in times of peace were for 5,000 troops.

The town is situated a mile to the west, with handsome wide streets, laid out at right angles. Within the last three years many public buildings and fine dwellings have been put up. The population is cosmopolitan, life reckless and picturesque. The summer heat is excessive and the winter sharp. Port Arthur is really a military and naval stronghold without trade. Its commercial complement is Dalny.

Life in Port Arthur is active, even strenuous, and impressed me the more after a view of the lethargy of Mukden, the expectancy of New-Chwang and the dreary panorama of mud-baked, silent, died-out villages through which I passed on my return ride around the Gulf of Pekin.

The streets are thronged with soldiers and companies of coolies who take a very serious view of life and do not hesitate to sweep the star-gazing stranger off his feet. You enter a street and suddenly the earth begins to tremble with the heavy tread of many soldiers. If you are wise you seek refuge in a Chinese house and then turn back to view the proud sight, for a proud sight it is, though often inconvenient.

Your Russian soldier is always on a war footing. If peace broods over the land he personally never recognizes it. He is always just going into action, grasping his rifle firmly, and with saber swinging free, singing as he marches a fierce, intimidating song. They stride along, red with the mud of the hills from which they come, after a few days in the field, with their faces haggard, their eyes deep set, and every man of them believes he is going under fire just around the corner, and as they approach the choruses swell in volume, the interruptions and the change of note from the leaders are more resonant, and you know without a word of Russian that they are spurring each other on to doughty deeds.

In Deadly Earnest.

They get around the corner, stack their arms, and are given something to eat, and at the same time one is impressed with the fact that your intelligent newspaper-reading soldier is a poor thing to look at in comparison with these disciplined savages. Your intelligent soldier looks embarrassed; he feels and looks as though he were masquerading, but your Russian is in deadly earnest.

Militarism is, of course, rampant, and but for a few Chinese who have found the Russian rule bearable, one seldom meets a man clad in anything but a uniform; the "rick-shaw," typical of the whole of China, has been placed by the "drosky" driven by "mujiks" of an unusually dirty type. As regards facilities for travel, there was (for it is wiser to speak in the past tense now) a splendid railway connection with the Trans-Siberian system, and on Mondays and Thursdays a through Pullman express ran to Irkutsk, meeting the transcontinental section from Vladivostok. A line of steamers under Russian control made daily trips to Che-Foo, eight hours distant, and Russian "tramps" had been taking more and more of the Korean coast trade away from the Japanese, who until recently possessed a practical monopoly.

Chain of Fortresses.

Forts of no mean kind and of great magnitude, can be counted, not in units, but by tens, between Dalny and Port Arthur. The industry displayed upon every hand in railroad construction, house building, the erection of fortifications, the making of docks, roads and the improvement of the harbor was admirable and commendable. All the works were pushed forward with perhaps ruthless but unflagging zeal and much prescience. Day and night operations have gone forward, designed to make Port Arthur a commercial emporium and a great naval arsenal. A newspaper printed in English, public water works, electric trams, electric lighting and much else—these are all upon the card, and were being got ready. At the same time the Russian government, as represented by Admiral Alexieff, had been feverishly busy laying mines, preparing the fleet for the war and searching for contractors who could and would deliver Cardiff coal in lots of 70,000 tons, less or more, up to 200,000 tons, early in 1904.

It was intended, once the harbor had been deepened over a greater area, to open a new channel, cutting this silted sand in a direction opposite the existing basin upon the far side of the water way. By that means the commercial marine would have its own part of the harbor and direct access to the traders' wharves and the new railway sidings. There is a rise of eight feet to twelve feet of tide at Port Arthur. The two largest battleships out from Europe found no difficulty in getting into the harbor, although they were said to draw over twenty-eight feet of water. They were at once taken into the basin, where they were touched up and painted in black within two days, like the other warships in port. For some mysterious reason the Russians divided their fleet, keeping the best part of their fast armored cruisers at Vladivostok.

Is Dirty, but Improving.

Port Arthur is, indeed, somewhat dirty and ill smelling. To the native Chinese smells are added the smell of moujiks, horses and leather; and while planning white Moorish hotels and macadamized avenues of acacias across the harbor, the Russians had been contented for five years with unpaved streets—mud sloughs, equalling those of Pekin in places. Since 1898 gold-laced uniforms have been bounced about in mud-bespattered droshkies.

While one finds every evidence of western civilization at Port Arthur, the native Chinese still cling to their simple ways. The flowing kimono, the tabi, or foot glove, and the geta, or graceful wooden sandal, are still seen. The visitor who is entertained at a native's home eats from an individual table about four and a half inches square, from an artistic lacquer bowl with either chop sticks or improvised spoons, and the hibachi is kept half full of ashes and burning charcoal for the social tea brew.

"Stop thief!" Great Britain or some other great world power cried every time Russia laid a new tie or pitched a Cossack tent on Manchurian soil. "Maintain the status quo!" they shrieked. "Keep the door open!"

Fine-sounding phrases! Lofty sentiments! Disinterested warnings! But they were belated.

CHAPTER XIII

RUSSIA'S CALL TO BATTLE

Czar Answers Mikado's Challenge to Combat and Army Reserves Are Called to the Colors—War Department Takes Charge of the Trans-Siberian Railway—Czarina Throws Kisses to Schoolboys and Students—People Kneel in the Snow Before a Chapel Containing a Sacred Image and Pray for Victory.

WILD scenes in the Russian capital followed the announcement that war had begun. Japan was denounced as a treacherous foe for having made her attack in the night. This fact inflamed the populace, and scenes of patriotism which accompanied the declaration of war against Turkey in 1877 were repeated.

In Kief, Odessa, Kharkoff, Ekaterinoslav and Moscow patriotic demonstrations were held. Public balls and other festivities were countermanded, and the Red Cross Society was besieged with women ready to go to the front as nurses.

The Czar's first move after accepting the issue of war was to order the mobilization of the army reserves in east Asia.

In every military district in European Russia regiments of infantry, cavalry and artillery were put under orders to prepare for a campaign in the far east.

The war department assumed control of the Trans-Siberian Railway, and its capacity was taxed to the utmost in the transportation of troops and munitions of war.

The state of feeling was illustrated at the theaters when people demanded the national anthem. More remarkable was the refusal of the drosky drivers to accept money from officers whom they drove to the palace.

There was a great scene at the naval academy when the Czar personally advanced the senior class to the rank of officers. The

Czar, who wore an admiral's uniform, in addressing the cadets said:

"You are aware, gentlemen, that war has been declared upon us. The insolent foe came by night and attacked our stronghold and fleet. Russia now needs her navy as well as her army. I have come today to promote you to the rank of midshipmen. I am confident that, like your revered predecessors, Admirals Chichagof, Lazaref, Nakhimof, Karnilof and Istomin, you will work for the welfare and glory of our beloved fatherland and devote all your energies to the fleet over which flies the flag of St. Andrew."

After his majesty's departure the newly promoted officers hired sleighs and drove up and down the quay fronting the winter palace, clad only in their black tunics, unmindful of the bitter cold, and shouting wild hurrahs. Grave visaged generals, carried away by emotion, saluted the youngsters, whose only regret was that their service uniforms were not ready so as to permit of their departure for the far east at once.

Orders Out 600,000 Troops.

On the second day after hostilities began the Czar had ordered an army of 600,000 men to be in readiness to resist the invasion of Manchuria by the Mikado's troops. The ukase, dated February 10, ordered all troops in the military district of Siberia to be placed in readiness for war, that all divisions in the far eastern viceroyalty be brought up to war strength, and that the army and navy reserves of the Siberian and Kazan districts be called to the colors. The military authorities were empowered to make requisition for the necessary horses.

There were six army corps in the far east, two each in the districts of Kazan, Siberia and Amur. Each army corps was made up of 1,030 officers, 47,653 men, 16,965 horses and 124 guns. The total strength of the six corps called into active service by the Czar thus approximates 300,000 men. The army reserves in the same districts doubled the force.

All men who had served five years in the army and were under forty-three years of age were required to report. This resulted in the disorganization of daily life in the empire, and had an ill effect upon every profession and calling.

The war fever, which was aroused all over the empire, or at least all over European Russia, in a single week was truly remarkable. Such patriotic demonstrations as were witnessed in St. Petersburg after hostilities began were all the more astonishing because of the public indifference during the period of negotiations and almost up to the hour when Japan struck the blow at Port Arthur.

But, like a flash, the whole empire seemed to have been lashed into fury by defeat, and, like a bear when it is wounded and angered, prepared to fight to the bitter end. The dreamy Russian character was stirred to depths of deepest resentment.

The patriotic demonstrations continued for three days and were entirely spontaneous. Some of them were extraordinary in character. The schoolboys and students of St. Petersburg, numbering thousands, marched bareheaded for hours up and down the Nevsky prospect, bearing flags, shouting and singing patriotic airs. They were followed by an enormous crowd.

Like previous processions, this one went first to the Anichkoff palace, where the Dowager Empress, who is very popular with the people, showed herself.

Later the boys and students completely surrounded the winter palace, and in response to their enthusiastic cheers the Czar and the Czarina made their appearance. When the Czarina blew kisses to the boys a vast shout went up which brought counter cheering from across the Neva.

At the French embassy, which was next visited by the procession, Mme. Pompard was compelled, in the absence of the Ambassador, to appear at a balcony and acknowledge the cheers for Russia's ally.

In front of the barracks the various regimental bands in the city played the national hymn to hurrahing crowds.

There were demonstrations outside the theaters, and the performances going on inside were stopped while the orchestras went out to the street to play the folksongs which the soldiers sang while marching.

Almost every town in European Russia had the same story to tell of popular demonstrations. At Moscow there were great manifestations in front of the palace of Grand Duke Serge and before the Iberian chapel, where stands the sacred image of Our Lady of Iberia, before which the Czar invokes blessings when he goes to Moscow. Thousands of persons knelt for hours in the snow in front of the chapel praying for victory.

Russian Peasants Give Savings.

In every sphere of life the Russians throughout the entire land taxed themselves to send money to one or another of the war funds. Even children emptied their savings boxes, and many cases were recorded where people in humble circumstances turned over the money saved in a lifetime to the Red Cross or to a patriotic fund of some kind.

At Blagovestchensk peasants gathered 400 rubles ($200) for the families of those called to the war, while a committee on the stock exchange at Moscow devoted 350,000 rubles ($175,000) to the same purpose. At Charkow the merchants organized a lazarette of 200 beds, while the Merchants' Club, in addition to large gifts of cash, agree that for every game of cards one ruble (50 cents) must be donated to the war fund, and for every game of billiards ten kopeks (5 cents).

The new director of the ministry of finance, M. Koscoutzoff, addressed the members of the stock exchange, impressing upon them their duty to be firm in this time of trouble and not allow funds and shares to be influenced by panic.

The most sacred image in Russia has been sent to St. Petersburg and it will be taken later to the far East with the army.

This image is a representation of the virgin appearing to St.

Courtesy of Everybody's Magazine.
ADMIRAL KABAYAMA.
Japanese Navy.

Courtesy of Everybody's Magazine.
MAJOR-GEN'L FUKUSHIMA.
Commander of the Japanese Forces in China.

Courtesy of Everybody's Magazine.
ADMIRAL ITO.
Japanese Navy.

ON THE MANCHURIAN BORDER.
A Trans-Baikal Railway and Military Station.

RUSSIAN FRONTIER POST IN MANCHURIA.
Russian Scouts Questioning Native Chinamen.

Sergius, and is always kept at the Troitzko monastery. It is about one foot square and is covered with precious stones. The image has a remarkable history. It accompanied Alexis, Peter the Great and Alexander I on all their campaigns. A silver tablet attached to it enumerates the battles at which the ikon was present.

The Czar's orders for the mobilization of a Russian army in Manchuria called to the colors all told about 550,000 men. As most of these had to be transported nearly 6,000 miles in wintry weather, the long delay between the outbreak of the war and the first big land engagement is easily accounted for.

Not since the Turko-Russian war of 1878-9 has any European power sent so large an army into the field. France fought Prussia in 1871 with 300,000 men. Russia sent scarcely 600,000 men against Turkey in 1878. England fought the recent Boer war with 200,000.

No power in Europe ever transported an army of 540,000 men so great a distance as Russia did in the conflict with Japan.

The first regiments to leave St. Petersburg for the scene of war were reviewed by the Czar. The following is characteristic of the addresses the Russian ruler made to his troops:

"My brothers, I am happy to be able to see you all before you leave, and I wish you a good journey. I am firmly convinced that you will all uphold the honor of your ancient regiment and readily risk your lives for your dear fatherland.

"Remember your foe is brave, confident, and crafty. From my heart I wish you success over your opponents.

"I bless you, my brothers. May St. Seraphim pray for you and accompany you in all your ways.

"I thank the officers for volunteering their services and once more I thank you all, my brothers, with all my heart. God bless you."

The troops then marched past, the Czar calling out to the men as they went by: "Good-by, my brothers."

CHAPTER XIV

HARBIN, RUSSIA'S ARMY BASE

Viceroy Alexieff Deserts Port Arthur a Week After Hostilities Begin and Establishes His Base of Military Operations at Harbin, 600 Miles North of the Besieged Fortress—Japanese Recognize the Change as a Shrewd Strategic Move—Description of the New Seat of Viceregal Power.

ONE week after the naval battle of Port Arthur, Viceroy Alexieff and his headquarters staff deserted the famous fortress and established a new base of military operations at Harbin, six hundred miles north of Port Arthur, where the Chinese Eastern railway joins the main stem of the Trans-Siberian line to Vladivostok. This move, which was mistakenly regarded as a sign of weakness on the part of the Russian commander, was in reality a wise strategic move. The Japanese commanders besieging Port Arthur by sea and land understood its significance.

Events had already determined that Russia could make no effective campaign on the water, and must rely upon her army to give her victories, if she was to be victorious. The army and its supplies had to be transported over the Trans-Siberian railroad, and the seat of the railroad management was Harbin. Inasmuch as the success of military movements and the necessity of maintaining connection with the home base of supplies, St. Petersburg, depended upon a single line of railroad, obviously the seat of military operations should be at the seat of railway operations, which in this instance happened to be the important junction point of Harbin, which was to assume new importance as Russia's capital in the far east.

Harbin is the most completely Russianized city in Manchuria. Possibly no other writer was so well qualified at the outbreak of the war to describe that really wonderful city as Henry B. Miller,

United States consul at New Chwang, who had made a careful study of conditions there, not only in the way of events but in their significance and their relation to American trade. What follows is from a report made by Consul Miller just prior to the war:

The Moscow of Asia.

"One of the greatest achievements in city construction that the world has ever witnessed is now going on in the heart of Manchuria.

"In the building of such cities as Vladivostok, Dalny and Port Arthur Russia has demonstrated her power and purpose on the Pacific in line with the world's conception of her character, but in the construction of this wonderful city of Harbin she is displaying an altogether different type of activity from what we are prone to attribute to her.

"It is in this city more than in all the others combined that Russia is asserting her intentions of becoming an active industrial force in the affairs of the orient, and her people are already giving the place the title of 'the Moscow of Asia.'

"The city is located on the Sungari river, at the point where the Manchurian branch of the Siberian railway crosses the stream and where the Chinese eastern branch starts south to Dalny and Port Arthur. It is about 350 miles west of Vladivostok and 600 miles north of Port Arthur. Its location is the geographical center of Manchuria, and from present prospects it is to become the commercial center as well. The city is surrounded on all sides for hundreds of miles with a rich and productive agricultural country, producing corn wheat, oats, barley, beans, millet, tobacco, hemp, vegetables and some fruits. Minerals and timber and great areas of grazing lands also surround it.

"At present the place consists of the old town, three miles from the central depot; Prestin, or the river town, the present commercial center, and the administration town, in close proximity to the railway station. Before the railway engineers established this as their headquarters there was no native town in this vicinity, and the entire place is therefore a Russian product.

For Russians and Chinese Only.

"It is as distinctly a Russian city as though it was located in the heart of Russia, and none but Russians and Chinese are permitted to own land, construct buildings or engage in any permanent enterprise. The city has been created by the Russian government, under the management of the Manchurian Railway Company. The land for many miles in each direction has been secured so as to make it impossible for any foreign influence to secure a profit or foothold close to the city, and foreigners are not recognized as having any rights whatever, but are permitted there by sufferance. The chief railway engineer is the administrator of the city, and up to the present time has had complete control of everything, but in the new scheme for the government of Manchuria some form of municipal organization will be permanently established.

"In 1900 the place began to assume importance as a center of railway management and in 1901 the population had grown to 12,000 Russians, in 1902 to 20,000; by May, 1903, to 44,000, and in October, 1903, a census showed a population of 60,000, exclusive of soldiers. Of these 400 are Japanese and 300 of all other nationalities, including Germans, Austrians, Greeks and Turks. All the rest are Russians. There are no Americans.

"The railway and administration employes, including families, constitute 11,000 of the population. The Chinese population is about 40,000, located in a special settlement. The ratio of women to men is as follows: Japanese, 20 per cent; Russians, 44 per cent; Chinese, 1.8 per cent; average of women, 14.3 per cent.

Center of Railway Control.

"Harbin is the center of the entire railway administration of Manchuria, and, as the Russian commercial enterprises of the far east are under the direction of the railway company, it will also be the center of Russian industrial and commercial development. It is the headquarters of the civil courts and the chief military post and the main center of control of all the vast army of railway guards. The administration city, therefore, consists of all of the public and private buildings and shops necessary for these various departments. Residences for the employes cover the largest area of this division of this marvelous city.

"The following are some of the principal buildings of the administration city:

	Cost.
Administration buildings, three stories in height, having a total floor space of 3,600 square sagene (176,400 square feet), to cost when finished	$ 618,000
Railway shops	1,287,500
Hospitals	322,300
Commercial school and girls' school	257,500
Technical school	128,750
Eight schools for teaching Russians Chinese and for teaching Chinese Russian	49,440
Club and store for employes	190,550
Hotel	83,945
Russo-Chinese bank	103,000

"The total administration expenditure on the city has been $15,450,000.

Excellent River Transportation.

"The Sungari river is navigable with light-draft steamers and native craft for nearly two hundred miles above the city, up both branches of the river, and much traffic has already developed on these streams, especially in wheat.

"From Harbin to the Amur river, during the navigating season, which begins in April and ends November 1, good-sized river steamers run daily. These steamers are well fitted with good, comfortable cabins for first, second and third class passengers.

They carry large cargoes of freight and usually tow barges loaded with freight. From Harbin to sea-going steamers at the mouth of the Amur cargo is carried now at about $4 gold per ton. The Chinese Eastern Railroad Company and the Amur Steamship Company run good steamers on this line, and there are also several private boats covering the same route. All are loaded continually to their full capacity.

"The steamers are mostly of the stern-wheel type, burning wood, such as are in operation on the western rivers in the United States, but as far as I could learn none is constructed of American-made machinery. The time usually required to go from Harbin to Harborofsk, at the mouth of the Ussuri river, on the Amur, is five days. At this place these steamers connect with trains for Vladivostok.

"Going west from Harbin the train takes you by a branch line from the crossing of the headwaters of the Amur to Stretensk, the head of navigation of this great river, while the main line goes to Lake Baikal (Siberia) and Russia. Going east, the railway reaches the sea at Vladivostok over a grade that does not exceed in any place thirteen feet to 1,000. Going south, the Chinese Eastern Railway meets sea-going ships at New-Chwang, Dalny and Port Arthur. The heaviest grade on this line is nine feet to 1,000, and that for only a short distance and at rare intervals.

"In October, 1903, the regular number of trains dispatched for through traffic was thirty per day. Eighteen local trains were dispatched in addition. These local trains connected the two extremes of the town, viz., the old town and Prestin, with the administration part of the city.

Electric Tramways and Automobiles.

"There are also about four hundred nesoshticks or Russian carriages for public use and the average earnings of these vehicles is $2.58 per day.

"There is also an automobile line ready to start four machines to operate between the old town and the administration city. Each vehicle will carry ten persons. These machines are now on the ground and will carry passengers for 10 cents each way. This line is in connection with an electric tramway that is to run a loop line through the river town, or Prestin, and a double loop or figure 8 line throughout the administration town. This is a private corporation, with a capital of $128,750. The same company is to provide an electric-light system for all three sections of the city.

"Harbin was started primarily as a military center and an administration town for the government and direction of railway affairs. Its growth into a splendid commercial and manufacturing city was not originally provided for by the promoters, and it has been somewhat of a surprise to them, but the fever of making it a great Russian commercial and manufacturing city has now taken possession of the railway management, and every system of promotion and protection that can be devised to increase its growth along these lines is being energetically encouraged.

Siberian Jews Supply Money for Enterprises.

"The capital for most of the private enterprises is furnished by Siberian Jews. Chinese are furnishing money for the construction of some of the finest private buildings, such as hotels, storerooms, etc. In the administration part of the city no private buildings of any kind are permitted.

"The old town was the first to be laid out and the land was sold to the public at the rate of 51.5 cents per forty-nine square feet the first year, but this rate is now increased to $1.55. Following this, in 1901, the administration town was laid out and construction work began on buildings covering 198,000 square feet. Later the river town, Prestin, was laid out, and in a very short time all of this was sold at a price of $8.70 per forty-nine square feet, and most of it is now covered with substantial brick structures, there being 850 buildings, constructed at a cost of $4,120,000. Recently two very large additions were laid out adjoining the administration town, and the land has been sold at prices ranging from \$2.57 to \$7.73 per forty-nine square feet. This was purchased largely by speculators and is being bought from them now at from \$10.30 to \$20.60 per forty-nine square feet.

"The administration has already received over \$1,030,000 for land sold to private parties. Many elegant residences and substantial structures are in course of construction in the additions adjacent to the administration town. A hotel and theater combined was built at a cost of \$30,000 and rented for \$12,875 per annum.

"All of this land is secured on an eighty-six years' lease.

Big Business of Russo-Chinese Bank.

"The Russo-Chinese bank is the only banking institution in the place and it has an elegant home in a structure of stone that has a steam-heating and electric-lighting plant of its own. The building cost \$103,000. The business of the bank has increased 30 per cent during the past year, and its daily transactions, exclusive of railway and other government accounts, amount to \$206,000. The bank makes no loans on realty, but advances from one-third to one-half capital for current substantial business. It is inaugurating a very efficient and active system of credits to Chinese merchants purchasing Russian goods for sale in Manchuria. In some cases as much as \$103,000 has been given in letters of credit to Chinese for purchases in Russia.

"These experiments are proving profitable and satisfactory. The largest success is reported in cotton goods. Many large orders are now being placed in this line, and a substantial trade is being created. These goods are brought into Manchuria via Vladivostok free of duty. So far sugar has been the only article purchased on which the Chinese have lost money.

"This system of advancing credit to Chinese merchants for the purchase of Russian goods prevails now generally throughout Manchuria, and it is by this method and by imports free of duty and favored rates over the railway that Russian cotton goods are likely to capture the great trade of Manchuria that is now largely in the hands of American manufacturers.

"The Russo-Chinese bank is also very generous to Chinese and Russian merchants in encouraging the purchase and shipping out of native products, but it is exclusive in its methods and will not encourage foreigners.

Milling Is the Leading Industry.

"The leading industry of Harbin is the manufacture of flour. Eight mills are now in operation, all with modern European machinery with one exception, and that is a small one constructed with American machinery. Applications have been made and granted for the construction of two more large ones, and by the middle of 1904 ten mills will be in operation, producing 902,800 pounds of flour per day. They pay from 30 to 35 cents gold per bushel for their wheat delivered at the mills, and the wheat-producing area can be increased enormously. The present value of the flour mills in Harbin is $618,000.

"In the immediate vicinity of Harbin there are 200 brick-making plants, the cost of which was $257,500. Two of these plants were constructed by the administration at a cost of $103,000. Most of the brick produced are used in the construction of the city. A very good grade of red brick is produced and sold for $3.35 per thousand. Most of the work is done by Chinese, who are paid 18 cents per day.

The Peoria of Manchuria.

"The next industry of importance is the production of the Russian liquor, vodka. There are eight manufactories, constructed at a cost of $103,000. Several of these produce vodka from spirits of wine and sugar brought from Russia; some produce only the spirits of wine from the local wheat, while others produce their spirits from local wheat and the vodka from their own manufacture of spirits. The consumption of vodka throughout Manchuria is something enormous. In Russia the production is very heavily taxed and it costs $5.15 per 2.707 gallons, while in Harbin it sells at from 77 cents to $1.28, this for 40 per cent alcohol. The bottles for this vodka are at present brought from Japan, but at Imonia—in Manchuria—the Russians are now building a large bottle and glass factory.

"Three breweries are now in course of construction in Harbin to cost $103,000. The Russians are great beer drinkers and produce very good beer, but it is not of the quality that bears shipping long distances, hence very little Russian beer is to be seen on the Pacific coast or anywhere in Manchuria. At the present time American beer has the best of the Manchurian market, as 150,000 dozen bottles are imported through one firm at Port Arthur every year. A fine quality of barley is produced in the Sungari valley, and these breweries will be able to buy it at about half the cost in the United States. There is little doubt that the Russians will soon be producing all of the beer consumed in Manchuria. Our Pacific coast hop men ought to be able to sell them their hops, however.

"These things, together with the financial help of the Russo-Chinese bank, have not yet been sufficient to do more than start the train well, and they may have to resort to a bounty in addition unless they can shut out foreign goods by a tariff.

"The production of cotton goods in Russia is growing very fast, and, as they have their designs on securing the trade in Manchuria in this line, it is only reasonable to suppose that they will eventually secure the trade they are prepared to handle in any country over which they have control.

"The following is from the pen of a well-known American writer, who has investigated the subject carefully and is thoroughly acquainted with the conditions of production and marketing of Russian cotton goods:

" 'There is considerable excitement just now about the Russian possession of Manchuria. * * * If Russia adopts the same methods as to other parts of Asia that she is now using in Persia she will drive all other countries out of the market. She has now the monopoly of the cotton business of Persia and she has gotten it by giving a bounty to her manufacturers. On every pound of Russian cotton goods sent to Persia the Moscow exporter gets an allowance of 3 cents from the government. One cent covers the freight and he gets 2 cents a pound profit, besides the usual profit on the goods. The English and German manufacturer has to pay full freight, with no rebate, and he can not compete. This same system will be adopted in China. * * * After the trade has been captured the rebate may be discontinued and the price will rise.'

Discriminate Against American Oil.

"Kerosene is the next in importance of American imports into Manchuria. Russian oil is already making very good headway in a free and equal competition with American oil. By forcing its use in all the cities of Manchuria, by special aid from the Russo-Chinese banks that are now established in all the principal cities, by preferential rates on the railway, by providing tank cars and tank stations along the railway line and refusing these advantages to American oil, it appears to me that Russian oil will have an absolute monopoly of the trade if full control of the country is secured to Russia.

"Concerning flour and lumber, I have recently issued detailed reports, the summary of which indicates that the Russians have it in their power not only to capture our trade in Manchuria, but to become our most severe competitors in all the oriental markets.

"In green and dried and canned fruit and vegetables I find the United States trade expanding considerably, and from every point of view within my observation I am induced to believe that the trade will have a large and permanent growth without danger of disastrous competition.

"Our trade in beer meets with the competition of Japanese and German beer, but it has been growing continually and is now greater than ever before. When the several breweries in course of construction at Harbin are in operation our trade in this line is most likely to suffer, and in time may give way entirely to the Russian product.

"In butter, the Siberian article is already capturing the Manchurian market, as it is being handled by the commercial department of the Chinese Eastern Railway. It may become a very aggressive competitor for the entire market of the orient. In condensed milk we have a large and growing market, not only in Manchuria, but throughout the orient. Should the Russian government elect to engage in this business also, it has the advantage of very cheap milk in Siberia and one of the finest countries in the world in the valley of the Liao, together with cheap labor to establish the industry on a basis that would make it a great rival for our condensed milk trade with Japan and China.

"The familiarity of the Chinese with hog-raising makes a good foundation for the growth of the industry, and I can see no reason why it should not continue to grow sufficiently to produce all that may be required for the oriental markets.

"There is a plant costing $12,875 for the preparation of bean oil for use in painting.

"Russians are especially fond of candies and sweets and few people know how to produce a quality equal to the Russian product. There is a manufactory in this line in the old town costing $5,150.

"There is on the river a small sawmill that cost $7,750 and two on the railway line between Harbin and Vladivostok that cost $17,500.

Agricultural Riches in the Vicinity of Harbin.

"There are many other industries in embryo, and, as the place is located in the center of an extremely rich agricultural country, has splendid transportation facilities and is doing so well in the establishment of manufacturing, there is little doubt that it will increase at a very rapid rate as a manufacturing and commercial center.

"The country is productive in wheat, cattle, sheep, hogs, millet, barley, oats, corn, beans, furs, hides, wool, bristles, bean oil, bean cake, hemp, tobacco and timber, and has various undeveloped mineral resources; in fact, it has all the natural elements for the foundation of a great city.

"The chief engineer who was in charge of the construction of the Russian railways in Manchuria informed me that Russia had expended in railways in Manchuria $139,050,000. Add to this her investments in fortifications and in the construction of the cities of Port Arthur, Dalny, Harbin and other places and it is a very moderate estimate to place her investments in permanent properties in Manchuria at a total of $257,500,000.

"What is the meaning to the United States of all this progress of Russia on the Pacific—the building of such cities as Harbin and the political domination of the country? It has been recently asserted by prominent people that it signified an enlargement of the market for our goods, and that of the presents imports into Manchuria 75 per cent were from the United States. General statements of this nature are easily made and easily believed, and without any careful examination into the details it has been the usual thing to assume that this development of Russia in Manchuria was certain to bring an increased market for the products of the United States. The subject has not yet been examined in all its phases as it should be, and, as far as I know, there is no one prepared by study and knowledge of all the details of the question to give a wise decision as to what the effect will be upon American trade from merely an economic point of view.

Manchuria's Imports from the United States.

"At the present time the principal imports from the United States into Manchuria stand in the following order: Cotton goods, kerosene, flour, lumber, canned and dried fruits and vegetables, beer, canned milk and butter, cigarettes and sundries.

"In cotton goods Russia is anxious for the trade and is making every effort to secure the business and is becoming a serious competitor. Her advantages in this line are political, bank advances and transportation. In a free contest, on purely economic lines, I think the United States can hold it. Russia favors the export of cotton goods into Persia by a heavy duty, and just what she will do in order to secure this trade in Manchuria is not yet determined. At present she is providing a heavily subsidized steamship line to bring these goods to Dalny and Vladivostok, where they enter free of duty, and no doubt they receive preferential railroad rates from these into the interior, or will if necessary.

"At Harbin an agent of a New York firm informed me that American trade there was confined now to canned goods, including fruits, vegetables, milk, etc., beer, sole leather, carts and a few lines of hardware.

"People informed me that they had succeeded in substituting Russian engines and railway material for American, and that the railway regulations now provided for the purchase of everything Russian, when possible, and that had cut off much American trade. They also said that they were succeeding in driving out American kerosene, flour, lumber, cotton goods and other things, and that they hoped soon to provide Manchuria with all the things that now come from the United States.

American Trade Better Under Chinese Rule.

"United States trade in Manchuria with the Chinese amounted to several millions of dollars per year and was almost entirely imports. It had grown very fast and would have had an extended and most substantial increase without the Russian development, for the country was being improved and extensively developed, with a continual immigration from other provinces in China, before the railway construction began.

"A study of conditions in Vladivostok, Harbin and other districts is not particularly encouraging to the idea of extension of American trade in Manchuria in any line that Russia is prepared to supply. A knowledge of the earnest intention of the Russo-Chinese bank to press the sale of Russian goods, a slight insight into the methods and determination of Russian railways to find a market for the products of Russia and the interest displayed in developing resources along their lines for Russians and Chinese only, taken in connection with the natural wealth and resources of the country, do not favor the hope that under a Russian regime our trade in Manchuria will be as large as it was before.

Greatest Problem in Asiatic Markets.

"If we take into further consideration the fact that the Russian government—by subsidies and through its banks and railways—is engaging in industrial and commercial pursuits as a government and calculate the cheap food, cheap and reliable labor, and the vast mineral resources that she will have at her command on the Pacific, the question of the Manchurian market becomes comparatively insignificant, and we find ourselves face to face with the greater problem of the markets of all Asia.

"With millions of cheap and efficient Chinese laborers, with vast coal fields bordering on the Pacific, with mountains of iron and copper, vast forests and enormous areas of agricultural land—producing now the cheapest food in the world—what is to prevent Russia, if her apparent plans are realized, from becoming a dominating factor in the commercial development of the far east? One cannot view the marvelous growth of a city like Harbin or observe the cities of Vladivostok, Dalny and Port Arthur and the great Siberian railway without pondering seriously the meaning of it all in the future of Russia on the Pacific.

"For the present the prospect is that we shall at least meet with such unfavorable conditions in Manchuria as will endanger our present lines of trade. Whether or not this will be compensated for by an increase in other lines is not at this time clear.

"There ought to be, and most likely will be, a large trade in agricultural implements. Of foreign countries Germany is securing the most of this trade now in Siberia and Manchuria, and there is no doubt whatever that German trade is benefiting enormously by Russian domination of Manchuria. Next to the Germans come the Austrians, and next to them the Danish.

Not an American in Harbin.

"It is not in the least inspiring for an American to go through as busy and active a trade city as Harbin and find so few things from his country and not a single American citizen or progressive business house. The vision of 75 per cent of American imports into Manchuria dwindles to a most insignificant amount. When you see the great flour mills continually enlarging and increasing in number, when you see the numerous breweries being constructed, when you see Russian engines and German, Austrian and Danish machinery and products and hear the successful development of Russian lumber mills and the introduction of Russian cotton goods, and see in the Chinese stores Russian oil and cigarettes where before were American, and where you hunt with straining eyes to find something from the United States, one is not seriously impressed with the statement that under Russian occupation our imports into Manchuria are sure to increase.

"Unfortunately the only customs returns by which we can measure our trade year by year in Manchuria are from the port of New-Chwang, and even that is very imperfect, for the imports all come from Japan, Hong-Kong and other Chinese ports, and the place of origin of the goods is not given in all cases. Goods are coming into Manchuria in great quantities through Port Arthur, Dalny and Vladivostok continually, as well as through New-Chwang, but there is no means of securing a proper report of them."

JAPANESE INFANTRY SCOUTING.
Mikado's Unmounted Troops Employing Method of United States Soldiers.

MARKET DAY IN A SIBERIAN TOWN.
Camels Are Used by Traders Because of Their Ability to Withstand Cold and Travel in the Snow.

RUSSIAN TROOPS IN CAMP.
A Scene Along the Line of the Terrible Winter March to Manchuria.

CHAPTER XV

RED CROSS IN THE WAR

Japan and Russia had Effective Organizations for Caring for the Sick and Wounded —Empress Dowager at the Head of the Society in Russia—Merchants and Churchmen Make Large Donations and American and English Women Aid in Preparing Materials—Japan at First Declines Outside Aid.

IN BOTH Russia and Japan the Red Cross Society has effective organizations, although the work of the Japanese corps seems to have been better systematized. The latter was so well equipped and prepared for hostilities that all offers of assistance from abroad were declined at the outset, much to the disappointment of many American physicians and nurses, who had expected to serve the little nation in its hospitals and on its battlefields.

In Russia the war plans of the Red Cross Society kept pace with those of the military departments. The head of the society in Russia is the Dowager Empress, the Czar's mother, a woman much beloved by the Russian people.

Count von Ontsoffdashkoff, one of the closest friends of Emperor Alexander III, was made the head of the Red Cross work under the Empress Dowager.

The Red Cross received an appropriation of $2,500,000 from the city of St. Petersburg at the very beginning, and the merchants of the city promised to give $750,000 more. The city of Moscow and the merchants there contributed $900,000, and the amount available for Red Cross work reached $5,000,000 within a week.

Niji Novgorod in a few minutes subscribed $150,000; Tula, $37,000; Tver, $12,500; Samara, $37,500; Rostowdon, $35,000.

The citizens of Paroslav presented the Grenadier regiment with the ikon of Holy Basil Constantin, the miracle worker, and with tea and sugar in great quantities.

At the famous imperial palace the hermitage was turned into

a grand central store, and everything needed in the care of the wounded and sick was gathered there in great quantities. It was a perfect museum of bandages, instruments and splints, and was presided over by nurses and doctors of the Red Cross.

Classes sprang up everywhere night and day, while armies of women were busied in preparing bandages and lint.

Dr. Carlovich, who had had much experience in the field, was the first to leave for the front with a staff of tried physicians. The Red Cross was overwhelmed with volunteers, and each section engaged to equip completely two hundred beds.

The holy synod gave $75,000 to the hospital fund and members gave up their salaries. The same was done by principals of many societies in order that the same might be devoted to the war funds. The Emperor's secretaries were kept busy thanking the donators, each one more liberal than the other.

The Dowager Empress issued a rescript to the Red Cross of Russia directing it to make provision to alleviate the sufferings of the wounded in the war with Japan.

American and English women in St. Petersburg started on the preparation of materials for the Red Cross Society with just as great heartiness as the Russian women engaged in the same work. Various committees and sewing circles, which had been already organized, effected a general organization at a meeting held in the Anglo-American church. These women realized that both armies engaged in the conflict would make large demands on humanitarianism, irrespective of race, and they naturally decided to assist the sufferers they could most easily reach—namely, the Russian wounded.

M. Alexandrovsky was placed in charge of the Red Cross field work and established his headquarters at Harbin. General hospitals were located along the Amur at Kharbarovsk, Blagovestchensk, etc., and the sick and wounded were transported to these points by the Sungari and Ussuri rivers and their tributaries, thus avoiding interference with the military traffic on the railroads.

One feature of Red Cross operations which had never before

been introduced during a war was the organization of small detachments to accompany the regiments into action. Each detachment consisted of two surgeons, four students and four body carriers. First aid to the wounded was applied on the battlefield.

Baron Korf, who was secretary to the St. Louis exposition commission, devoted his entire time to Red Cross work as chief of the personnel division. When explaining the extensive preparations of the society, Baron Korf said: "We are preparing for a long, hard war."

Almost all the private concerns at St. Petersburg and Moscow and other large cities which intended sending exhibits to the St. Louis exposition decided not to do so, and donated the moneys they would have expended in this work to the Red Cross Society and other war purposes.

The devotion of the young Empress Alix to the work of the Red Cross Society aroused the enthusiastic admiration of the Russian people.

She threw herself heart and soul into the work, even personally assisting in the preparation of bandages and other supplies for the sick and wounded.

As an evidence of the interest and activity of the Empress, she frequently started work at 8 o'clock in the morning and did not stop until 3 o'clock in the afternoon, not even taking time for luncheon.

Her little daughters strove to emulate the example of their mother and assist in collecting and packing the materials.

Miss Clara Barton, president of the American Red Cross, called at the Russian embassy in Washington soon after the beginning of hostilities and offered the services of the society to the Russian government. The ambassador expressed the thanks of his government on the kindly offer, but explained that outside aid would not be needed in view of the effective work of the Russian Red Cross Society.

Japan also refused many offers of Red Cross aid from America. The famous Chicago surgeon, Dr. Nicholas Senn, a great admirer of the Japanese, had promised the war officials of Japan to give them his services whenever they should be needed, and when hostilities opened Dr. Senn was ready and willing to go, but was not called upon. The Japanese Red Cross Society is organized upon the same lines as the American Red Cross and was splendidly equipped in every particular. The medical staff were men who ranked high in their profession. The Japanese Red Cross proved itself remarkably efficient in field work as well as in the conduct of permanent hospitals.

Although the aid of several noted surgeons who volunteered was not needed, Japan accepted the services of some American nurses, notably a party headed by Dr. Anita Newcomb McGee of Philadelphia. The other members were Misses Ella V. King, Minnie Cooke, Adelaide Mackereth, Elizabeth Kratz, Adele Neeb of Philadelphia; Miss Sophia Newall, New Jersey; Miss Genevieve Russell, New York City; Miss Mary Gladwin, Boston; and Miss Alice Kemmer of Indiana.

The Russian field corps of the Red Cross Society was amply equipped, even for the most unusual emergencies—one novel feature being a large number of dogs trained to the work of finding wounded soldiers and carrying with them the bandages and other supplies generally known as "first aid to the injured."

CHAPTER XVI

THE PRESS AND THE WAR

How the News of the War Reaches America After Traversing Fifteen Thousand Miles of Cable on the Bed of the Ocean—Cablegrams From China Cost the American Press 38 Cents a Word and From Japan 50 Cents a Word—Route of a Press Dispatch From Nagasaki to Chicago.

NEWS of Japan's victory at Port Arthur reached the United States after having traveled fifteen thousand miles of cable on the bed of the ocean. Much of the direct news originated in southeastern China, from which points the rate was 38 cents a word, while 50 cents was charged from points in Japan. These tolls were a special rate for press dispatches only, the regular commercial rate being three times as large.

The tiny electric impetus put in motion by the key of the operator in far Nagasaki instantly plunges under the eastern sea and comes to land in China, near Shanghai, 476 miles away. Then that little throb works southward round the China coast to Hong-Kong, 945 miles. At Hong-Kong (British) it dives under the China sea to Saigon, in Anam (French), 951 miles; from Saigon it crosses the bed of the sea to Singapore (British), 626 miles. Or it goes by way of Labuan, Borneo (British), 1,971 miles.

Through the Malacca strait to Penang (398 miles), and then a great plunge westward through the wild Nicobars and under the tropic Bengal sea (1,389 miles) to Madras. At Madras it takes to the land until it comes to Bombay.

Never resting, the brave little spark takes to the water again, traversing the broad Arabian sea to Aden (1,850 miles), threading its way up the scorching Red sea, flying ever westward, to Alexandria (1,534 miles). And from Alexandria it travels deep under the balmy Mediterranean to Malta, out to Lisbon, and so to London (3,205 miles), and thence across the Atlantic.

Every word forced so laboriously through those 15,000 miles of solid wire cost 50 cents.

It is difficult to give any very definite idea of what war costs the newspapers for telegraphy. Probably few persons have any idea of the vast sums which are swallowed up in a single day's news provided by the New York and Chicago newspapers on the occasion of, say, a great naval battle.

For two short messages from Japan, consisting of about 100 and 150 words, respectively, the bare telegraphic cost approaches $150, although some of the "the's" and "and's" in such cablegrams as they appear in the newspapers would not be telegraphed. In the event of a big battle, the big American papers pay thousands of dollars in cable tolls alone.

All the telegraph lines in Japan are owned by the Japanese government, and censorship of messages was therefore, easy. All messages sent by war correspondents had to be edited by the press censor who cut out any words or sentences he deemed objectionable. This was true of messages either from Russia, Japan, Korea and Manchuria.

Moreover, for the first few weeks press messages handed in in Japan had to be paid for in cash. A newspaper correspondent had, therefore, to provide himself with large sums of money, which was often extremely inconvenient. Afterwards the Japanese government conceded that point, and allowed the telegraph companies to take the risk of payment upon themselves and collect the money from the headquarters of the newspapers on this side.

At Nagasaki, the "taking-off" point for the mainland, messages were transferred from the Japanese government lines to the Great Northern company (Danish), and crossed either to Shanghai or Vladivostok. From Vladivostok the Northern company's line follows the railway track across frozen Siberia to Libau, on the Baltic. But few of the American press messages took that course, although there was in existence a sort of promise by Russia that no messages should be interfered with.

At Shanghai begins the cable of the Eastern Extension company, and the eastern cable takes up the thread at Bombay. From Bombay, also, the Indo-European line starts away and travels overland by Bushire and Teheran, Tiflis, Odessa, and Warsaw to Berlin, and so to England.

The American Commercial Pacific cable goes to the Philippines, and does not touch Japan. The possible routes for the direct transmission of news from Japan were, therefore, but two in number—the overland Russia route and the coast route via India and the Mediterranean or India and Europe.

The following table shows the route taken by messages sent from Nagasaki and Che-Foo, indicating the relay points:

	Miles.
Shanghai, eastern sea cable	476
Hong-Kong, Chinese telegraph	945
Manila, Commercial Pacific cable	729
Guam, Commercial Pacific	1,709
Midway Islands, Commercial Pacific cable	2,693
Hawaii, Commercial Pacific cable	1,384
San Francisco, Commercial Pacific cable	2,412
Chicago, overland telegraph	2,260
Total	12,608

Messages sent west from Japan or China had a choice of several routes, especially after leaving Lisbon, when they could go either via the Azores or via London and Waterville, Ireland. The route which was followed in sending the message around the world at the opening of the Pacific cable last July and the lines and distances composing it from Nagasaki or Che-Foo follow:

	Miles.
Shanghai, China, eastern sea cable	476
Hong-Kong, Chinese telegraph	945
Saigon, Annam, China sea cable	951
Singapore, China sea cable	626

Penang, cable Malay peninsula	398
Madras, Bengal sea cable	1,309
Bombay, overland telegraph	675
Aden, Arabian sea cable	1,850
Alexandria, cable and overland	1,534
Malta, Mediterranean cable	913
Gibraltar, Mediterranean cable	1,126
Lisbon, cable	392
Azores, Atlantic cable	1,053
Canso, Atlantic cable	1,698
New York, Atlantic cable	893
Chicago, overland telegraph	990
Total	15,909

At the very outset of the war the general manager of the American Associated Press made a direct appeal to the Czar to remove the press censorship on dispatches from Russia, which was done, and such war news as reached St. Petersburg was given to the world freely.

Japan, on the other hand, established a rigid censorship and no dispatches were permitted to be sent from Japanese points until they had been passed upon by a Japanese official. As Japan's policy was to keep the movements of her land and naval forces a secret, much news was suppressed by the Japanese censor and much was sent out that was not true. A large party of American and English war correspondents who went to Tokio at the beginning of the war were detained there for several months by the authorities.

CHAPTER XVII

TORPEDO ATTACK PROVED EFFECTIVE

The Torpedo an American Invention Which Has Been Most Highly Developed in Austria—Torpedo Boats Compared to Battleships and Cruisers—Daring Life on a Destroyer—The Stiletto of the Navy With Which Deadly Blows Are Struck in the Dark—The Whitehead Torpedo and How It Is Launched.

AN AMERICAN first suggested the idea of the torpedo in naval warfare, and American vessels first demonstrated the feasibility and effectiveness of this form of attack during the American Civil War. The first machine of the kind was the invention of David Bushnell, an American Revolutionary War patriot, but the attempt to use it upon a British ship in New York harbor was a failure because the so-called "torpedo" could not be attached to the vessel and it escaped. In the American Civil War the torpedoes were carried on the end of a spar and were ignited by electricity.

At the very beginning of that war the first automatic torpedo was invented by Captain Lupuis, an Austrian naval officer. From that day to this no nation has demonstrated so clearly the effectiveness of the automatic torpedo as the Japanese. Although the United States was the first government to introduce this powerful, mysterious and truly terrible weapon of modern naval warfare, she has practically ceased adding to her flotilla of torpedo-boats. Generally speaking, it has been believed that the peculiar opportunities for the effective use of the torpedo-boat were so rare as to make it a negligible quantity in modern war, but the deadly work accomplished by the Japanese torpedo boats has changed this opinion, and the torpedo attack is again recognized as a method of warfare to be reckoned with in all naval operations in war.

The Weak Spot of Battleships.

In the first Port Arthur engagement the Cesarevitch and the Retvizan each displaced about 13,000 tons, and the Palloda a little less than 11,000. It is doubtful if the torpedo boats which successfully attacked them displaced more than 350 tons each, if they did that. Their vanquished foes were at least thirty times as big, but, like Achilles, battleships have their vulnerable parts. Hit them below the armor belt and they are disabled, if not wrecked.

Naval experts have long had fears on this score, which they were not fully able to confirm or suppress. With the exception of a brilliant exploit in the harbor of Wei-Hai-Wei in February, 1895, there never has been a practical test of this kind of warfare until now. Japanese torpedo boats at that time sank three Chinese war vessels—the Ting Yuen, Chin Yuen, and Lai Yuen. Since China did not rate as a first-class military power, however, it was not felt that the test was significant. Both the United States and Spain had thoroughly modern torpedo boats in Cuban waters in 1898, but neither of them experimented with the missiles for whose use they were designed. Nothing further was learned, therefore, until the successful torpedo attack upon the Russian squadron at Port Arthur.

The eyes of naval officers all over the world were turned toward the oriental waters to watch the movements of the torpedo boats, and their stock in the estimation of many naval men who have heretofore considered them more spectacular than useful.

A torpedo boat, compared with a battleship or a cruiser, is a cheap boat. To be effective they have to attack in swarms, for the single torpedo boat that would dare to approach a hostile fleet would most probably never live to launch its torpedo. Being quick, cheap and carrying a small crew, if several are lost and several costly battleships of the enemy destroyed or disabled, naval men argue, they are economical.

Torpedo Boat a Weapon of Darkness.

One reason for the success of the Japanese is to be looked for in the time selected to make their attack. By the very nature of things the torpedo boat is the weapon of darkness. It is the naval stiletto, the blade that moves swiftly and secretly. It is too fragile to expose itself to the fire of a ship's guns. As a matter of fact, small-arm fire is sufficient to disable and sink one of these almond-shelled craft if delivered, say, from a machine gun.

Night, then, is pre-eminently the time for a torpedo-boat attack. The history of the previous attempts shows that all the successful attempts yet made have been made at night, and, on the other hand, to emphasize this statement, every day attack has failed and frequently brought disaster upon the little boats. Having had experience that no other nation has had, it is not remarkable that the Japanese knew how to make such an attack effective. They have also once again repeated the advice of the theorists and have proven that theory to be the correct one.

The torpedo boat itself is the veriest cockleshell afloat. Its plates are not any thicker than the binding boards of an ordinary book. When you have a stretch of these plates over 150 or 200 feet they naturally become weakened. Everything is speed. Give the torpedo boat thirty or thirty-five knots and the governments do not care what her plates are so long as they will keep her boilers and machinery in place and provide a fairly good gun platform. The quarters for men and officers are cramped, for room is needed for boilers and coal.

Life on a Torpedo Boat.

Life on a torpedo boat is exciting, even in times of peace, for a sea wave may dent in or crumple up her nose or even sink the craft. In war no insurance company ever conceived would write a policy for any member of the crew. When the torpedo boat darts out on her attack every soul on board literally takes his life in his hand. The dangers are at least in the ratio of ten to one of those on a battleship.

To steer a torpedo boat into a fight requires a steady, cool and fearless head. The torpedo, which weighs half a ton, must be adjusted for distance, put into the launching tube, and then the boat must be quickly maneuvered until it is at the right place.

This is the psychological moment. While the boat is running swiftly the torpedo must be launched, and so accurate are the torpedo adjustments that it will infallibly go just as directed, and even such deviation as may be given it by a rope or obstruction is automatically overcome, and it resume the original course planned for it.

Development of the Torpedo.

The history of the torpedo and its development is both interesting and important to all readers of naval warfare. The first navigable torpedo was a cigar-shaped metallic object containing machinery for its propulsion and a charge of some explosive at the front end. In some of the early designs an electric motor drove the screw, the current being supplied through a wire from the shore. This wire would be previously coiled up inside the torpedo and stretch out when the latter was moving. The rudder could be controlled from land, too, by electrical means, and a deviation from the original course amounting to 30 or 40 degrees was thus made practicable. In the Sims-Edison design the depth of submergence was regulated by suspending front and rear from a tiny float. Several inventors proposed to make the torpedo entirely independent of land by supplying motive power of another kind. One man stored a quantity of compressed carbonic acid there, and used it to actuate an engine.

Another introduced a heavy flywheel, to which, just before launching, it was proposed to impart a high rotative velocity by outside mechanism. Fully 10,000 revolutions a minute could be secured. In that way enough power was to be stored to drive the torpedo a few hundred feet.

The torpedo invented by Captain Lupuis, previously referred to, was run by clockwork and guided from shores by ropes. The government liked the idea, but recommended the selection of a better motive power and a simpler means of guiding. Three years later Lupuis met Whitehead, then manager of an engine manufacturing company at Fiume, and exhibited his torpedo plans. Whitehead, with the assistance of his son and a skilled mechanic, secretly made the first Whitehead torpedo and two years later submitted it to the Austrian government. Externally it had the appearance of a modern torpedo; its weight was 300 pounds and it carried a charge of eighteen pounds of dynamite. A compressed air chamber, charged to a pressure of 700 pounds to the square inch, supplied the motive power. For short distance the torpedo attained a speed of six knots.

The Famous Whitehead Torpedo.

The Austrian ordnance officers were enthusiastic over the torpedo, although it was a very crude affair. The government was too poor to buy it outright, but paid for the right to make the torpedo after Whitehead's plans. England a year later secured the right to make the torpedo. France, Italy and Germany followed. England has manufactured the Whitehead on a large scale at the royal laboratory, at Woolwich and at the Whitehead factory at Portland, about 6,000 having been issued to the navy.

In 1897 the manufacturing of torpedoes in England was abandoned and the English admiralty began to import its torpedoes from the two European factories, one in Germany and the other in Fiume, Austria. England's giving up of home manufacture is said to have been the result of the inferiority of English-made torpedoes to those made on the continent and in America. Torpedoes for the American navy are built in Brooklyn by the government, under a license from the patentees.

The American Whitehead not only has the power to blow up any ship afloat, but its intricate and delicate mechanism makes certain its path under the water. The variations from its course are so slight that it can be fired from the launching tube with the same confidence in its ability to reach the target as when the seacoast artilleryman fires a steel shell from a heavy gun.

The torpedo is built of steel in the shape of a porpoise, with a big double-bladed tail. Ready for firing it weighs 1,160 pounds, but its weight in water is but a half pound. Its length is five meters (about sixteen feet five inches), its greatest diameter forty-five centimeters (17.7 inches). The walls are made of the finest forged steel, to resist the enormous air pressure. Bronze bulkheads separate the sections.

Launched By Compressed Air.

Compressed air is the motive power. This is contained within the air flask, a hollow forged steel cylinder, nearly half as long as the torpedo, slightly tapering at the ends, with dome-shaped heads screwed and soldered in each end. On shipboard this flask is filled by an air-compressing engine, and the pressure attained is 1,350 pounds to the square inch. The flask is tested for a pressure of 2,000 pounds. This great pressure so compresses the air that the weight of the ten cubic feet in the flask is 69.19 pounds. A pipe connects the flask with the engine-room, a small compartment in the forward part of the after body.

The engine consists of three cylinders radiating out from the propeller shaft like a three-leaf clover. The cylinders could be carried in one's overcoat pocket, but they have a combined power of thirty horse power. A main crank, turning the propeller shaft, receives its impulse from the piston of each cylinder in succession. Three slide valves on each cylinder regulate the admission of air. The propeller shaft turns two double-bladed screws in the tail in opposite directions, a gearing in the after body giving the reverse motion to one propeller. The two propellers neutralize their individual tendencies to cause the torpedo to roll. To secure the right balance between them the propellers are shaved down after experimental runs.

The form of tube now being issued to the navy for broadside firing is about thirty feet long. The lower section of the forward half is cut away, leaving a long, overhanging spoon sticking out. Inside the cylinder, on the top, is a T slot, extending from the breech to the end of the spoon. In this T slot fits a small T projection on the top of the torpedo. Suspended by this projection the torpedo balances, so as it is sent out of the tube the T carries it out to the end of the spoon in a horizontal position, and, slipping out of the slot, the torpedo strikes the water nearly level. Without the suspension arrangement the nose of the torpedo would dive down as it slipped over the forward edge of the tube, resulting in a deep initial dive.

Big Warships Carry Torpedo Tubes.

Nowadays nearly all big warships have three or four torpedo tubes. For real work much smaller and much faster craft are needed. The size of torpedo boats has undergone some curious changes. Their original function—theoretically, at least—was coast defense. It was hardly thought that they would ever be employed outside the harbors of the country which owned them. They were meant to resist invaders, not for aggression. They had a displacement of only twenty-five or thirty tons, and could have been hoisted up to the deck of a battleship or cruiser, to be taken to a distance, if necessary. Then came a rapid development in proportions. The Vesuvius, of the American navy, which represents the influence of that movement, displaces 930 tons. England projected several boats of from 800 to 1,000 tons each. At length however, a reaction set in, and to-day the limit is about one-quarter or one-third of the latter figure. The Dupont (American) displaces 185 tons, and the Sokol (Russian) 240, while the ill-fated Viper (British) had a displacement of 370 tons.

The greatest advances, however, are in sea-going qualities, armament and speed. The best torpedo boats to-day have guns that will enable them to sink unarmed vessels of that class, and are able to cross the ocean. Practically all torpedo service now is performed by "destroyers." These are adapted to making 25 to 30 knots an hour, whereas the earlier torpedo boats were good for only about 20. The Vesuvius is credited with 22½, the Dupont with 28, the Daring and Havoc (British) with from 28 to 29. The Sokol made 30.3 knots on her trial trip; between 30 and 32 knots was expected from the last batch of Japanese destroyers, nineteen in number, whereas the Viper, equipped with Parsons engines, made 32 knots before her accidental destruction.

Speed at High Cost.

Speeds like these enable boats to make sudden dashes and retreats, but they are obtained only at a high cost. Battleships, the slowest of all naval vessels, have scarcely more than one horse-power to a ton of displacement. The "crack" battleship of the American navy, the Missouri, displaces 12,230 tons, and her engines have 16,000 horse-power. The latter figures indicate the capacity of the Retvizan's engines, and her displacement was 12,900 tons. Now, the Cushing has a displacement of 185 tons, and a horse-power of 3,500 at her command. The same ratio—about one to eighteen—is furnished between the tonnage and power of the Sokol. In the Daring difference is even more marked. She displaces only 220 tons, and developed 4,735 horse-power on her trial. The Viper displaced 370 tons, but her engines showed close to 11,000 horse-power!

The newest type of torpedo boat is the submarine. This style of vessel runs on the surface until within a mile or two of the enemy, then becomes partially or entirely submerged, and completely disappears from sight just before discharging a projectile. So long as its smokestack can be allowed to stick up out of water the boat is propelled by gasoline engines. When she dives the screws are driven by storage batteries and an electric motor. Even on the surface these boats run slowly, none yet built making more than ten or eleven knots, or traveling faster than six or eight when fully immersed. They are also much smaller than the average destroyer. Boats of this kind are provided with additional mechanism to maintain a level course under the surface, to take in and eject water rapidly and for making observations when the hull is completely out of sight. For the last purpose an optical instrument of peculiar construction is mounted at the very top of

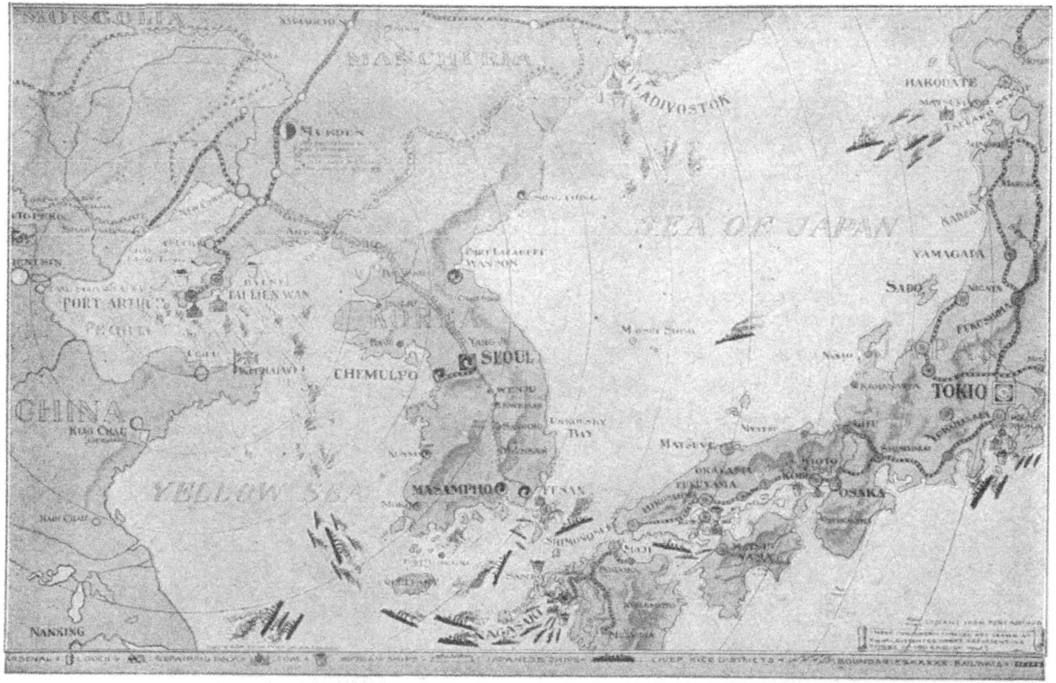

NAVAL STATUS OF JAPAN AND RUSSIA AT THE OPENING OF THE WAR.

THE above map shows the position of the Japanese and Russian fleets at the opening of the war, with the location of repairing docks and coaling stations, arsenals and railways. The chief rice districts are also indicated, and the map is both artistic and complete.

THE MOLOCH OF WAR IN THE FAR EAST.

JAP and Russ are battling ferociously around their tattered standards of war. In the background is an allegorical figure of War, casting forth forked lightning from one hand, and with the other waving the flaming torch over the devastated land. It is War in terrible allegory.

the tube, which stands up like a tiny smokestack, and in which mirrors or prisms transmit the picture to an observer inside the boat. The method of discharging a torpedo is the same with a submarine as with any other torpedo boat.

France and the United States have been pioneers in this line of experiment. England, Germany, Italy, and even Sweden, have followed suit. Russia has also conducted experiments, and in 1902 and 1903 had seven boats under construction. So far as is positively known, Japan had no contracts out for submarines, but she may have quietly purchased a boat or two.

Lieutenant Commander Frank F. Fletcher, U. S. N., commandant of the United States naval torpedo station at Newport, whose inventions of torpedo appliances and long experience with these dangerous projectiles have caused him to become recognized as one of the ablest torpedo officers in the service of the United States, is an enthusiast in the matter of the efficiency of torpedoes in warfare, and a lecture he delivered at the United States naval war college upon the history of the use of torpedoes gives a record of every attack made in the world with torpedoes from the time of the Civil War through the Russo-Turkish campaigns, during which the old spar torpedo gave way to the present automobile, down to the very beginning of the war in the Far East. It is, perhaps, the only record of its kind in existence.

There are records of fifteen attacks with the spar torpedo, which were attached to fast launches by means of a spar. Lieutenant Commander Fletcher says a good percentage of these attacks were successful. Nine per cent of the boats making the attack were lost and 3 per cent of the men were killed. On the other hand, six ships were sunk and three damaged, while 500 lives were lost. Sixteen torpedoes were exploded 45 per cent of which proved fatal to the ships.

With the automobile torpedo nine attacks have been made, and Lieutenant Commander Fletcher has summarized them as follows: Five hundred men took part, and the loss of life was only 2 per cent. Thirty-two torpedoes were discharged, and nine of them made hits, sinking eight vessels, the per cent of hits being over 28 per cent. "Thus," says Lieutenant Commander Fletcher, "the crucial test of war shows that the torpedo within its range is more accurate than the gun in battle."

Spar Torpedoes.

Boat attacks with spar torpedoes, which were started in the Civil War, covered a period of twenty years, and were employed in four wars. Attacks with the automobile torpedo have covered about the same period, and have been also employed in wars in which seven nations have been engaged.

The first attack with a spar torpedo was made against the Ironsides at 9 p. m. October 5, 1863, while the vessel was lying at anchor off Charleston. The Ironsides was severely injured, but not to such an extent that she had to be withdrawn from service.

The second attack was made on the Housatonic four months later, also off Charleston. It was moderately dark and the Housatonic was sunk with a loss of five lives. The torpedo boat was sunk with the loss of nine lives.

The third attack was made against the Memphis in the North Edisto river at 1 a. m. March 6, 1864. The attack failed and the torpedo boat escaped.

The fourth attack was made upon the Minnesota, off Newport News, April, 1864. Although the boat had been seen, she succeeded in exploding her torpedo amidships and doing much damage. The boat and her crew escaped.

The fifth attack was made upon the Wabash on blockade off Charleston in April, 1864. The attack was discovered and abandoned.

Lieutenant Cushing's Feat.

The sixth attack was that on the Albemarle, in October, 1864, the vessel being moored in the Roanoke. It was the event that made Lieutenant Cushing famous. The torpedo boat was discovered, but she pushed on through a very severe fire. The torpedo was successfully exploded against the ship, which was sunk. Of the crew of the torpedo boat only two lost their lives, being drowned.

The seventh attack was made by the Russians against Turkish vessels in the Black Sea, in May, 1877. Boats fitted with spar torpedoes failed to do any damage and escaped without loss.

In the eighth attack, which was made in the Danube, in May, 1877, one vessel was sunk by four launches.

In the ninth attack, which was made off the mouth of the Danube, in June, 1877, although participated in by five Russian launches, the ships of the enemy escaped damage.

The tenth attack also occurred in the Danube in June, 1877, and was also a failure. It was attempted in daylight.

The eleventh attack was also a broad daylight affair. It was made by two boats against a Turkish monitor in the Danube, and was a complete failure.

The twelfth attack took place in the Black Sea, in August, 1877, against a Turkish fleet, which escaped injury. The boats, though, got within torpedo range and there was no good reason why ships were not sunk.

The thirteenth attack was made by a force of Chile against a Peruvian vessel in the harbor of Callao. It was unsuccessful.

The fourteenth attack with spar torpedoes was the most successful in the history of such appliances. It was made by a French force against a Chinese vessel in the harbor of Foochow. Two boats participated and the man-of-war was sunk and 255 were killed.

The fifteenth attack was made by the force of Chile against a Peruvian vessel in Sheipoo, in February, 1885. It was made under great difficulties, many precautions against attack having been taken. One man-of-war was sunk and later the other foundered, having been damaged, it is thought, by the guns of her consort.

First Use of Automobile Torpedo.

The Russians first used the automobile torpedo, a Whitehead, in December, 1877, when it was a very crude affair. It was directed against a Turkish fleet at Batum, but failed to hit.

The second attack with the Whitehead torpedo was made by the Russians at the same place a month later. The Turkish guard-ship was sunk by two torpedo boats, which thus recorded the first hits with the automobile torpedoes.

The third attack, which was unsuccessful, was made against a Chilean ship in the harbor of Valparaiso in January, 1891.

In the fourth attack the insurgent Chileans sank the government ship Blanco Encalada in Caldera Bay in April, 1893. Two hundred of the crew were lost. The two torpedo boats which came from Valparaiso, many miles away, escaped.

The fifth attack in which automobile torpedoes were used was that in which the Aquidaban was sunk in Santa Catherine Bay in April, 1893. Two boats having failed made a second trial the next night. The first torpedo failing to make a hit a second was deliberately discharged in a rain of bullets from the ship.

The most desperate attacks with torpedo boats and automobile torpedoes were made against the Chinese by the Japanese. The first of these was at the battle of Yalu, and, the aim being poor, the attack was unsuccessful.

The next attack with torpedoes was made by the Japanese against the land forces at Port Arthur merely to create a diversion.

The eighth attack was made by the Japanese at Wei-hei-Wei, a whole fleet of Chinese vessels being anchored in the harbor at the time. The fleet had taken every precaution, but one ironclad was sunk and one torpedo boat lost.

The ninth attack was made in the same place on the following night, four boats composing the offensive force. In all eight torpedoes were discharged and two cruisers and two other boats were sunk.

CHAPTER XVIII

VLADIVOSTOK, THE PIONEER OUTPOST

First Russian Stronghold Built in the Far East—Entrance by the Golden Horn like the Golden Gate of San Francisco—Life in the Remote Military and Naval City which is a Cheap Imitation of St. Petersburg—A Fortress as Impregnable as Millions of Dollars Combined with Science and Nature Could Make It.

VLADIVOSTOK, the eastern terminus of the Trans-Siberian railroad on the Japan sea, was made as impregnable as the expenditure of millions of dollars could do, combined with science and nature. The harbor is admirably located, with narrow entrances and surrounded on all sides with hills which give good elevation to batteries. The entrance from Peter the Great bay resembles that of San Francisco; they are almost identical in conformation and also in names—The Golden Gate of San Francisco and the Golden Horn of Vladivostok.

The city is almost at the toe of the Muravey Amurski peninsula on a harbor which affords excellent anchorage for warships as well as merchantmen. A large navy yard with dry docks and repair shops is located here. The entrance to the harbor from the outlets to the sea are guarded by heavy batteries.

Two years before the war a ship canal was built which permits torpedo craft from the defenses to make a sortie against an enemy without attempting to run the batteries.

The city is well protected against land attack by strings of land batteries connected by sunken roads.

The first Russian stronghold constructed in the far East, Vladivostok stands as the pioneer outpost of Russian advance into that region. Situated on a gentle slope of hills, on a curve of the Golden Horn, the city is so screened as to be entirely concealed from without the harbor. Owing to the strategical position of the many small islands lying in and about the harbor, heavy pieces of ordnance were mounted on commanding and advantageous points to insure safety from a sea attack. There are, all told, forty-three fortified positions protecting the harbor entrance.

Vladivostok (pronounced Vlad-e-vo-stock), is essentially a port of and for the Russians. This is manifested in a marked degree by the absence of English speaking clerks in the hotels and commercial houses, which is not the case in any other part of the orient. The Russians do not encourage immigration of aliens, but supplement the population by sending numbers of their own people out, either by steamer or across Siberia by the great railroad.

Military City.

Vladivostok is a military city. Commercial houses thrive there, Asiatic aliens settled in great numbers, but above the clattering of commerce, the confusion of tongues, the click of machinery, can always be heard the voice of the military. The houses of the lower classes are particularly squalid and dirty, justly fitted to shelter those whose ideas on cleanliness are in strict accord with the condition of their abodes. The streets are like the houses, dirty, dusty and the receptacle for the garbage of the poorer classes.

A drive over the roads of Vladivostok is both instructive and exciting and affords one a sufficiency of exercise that even several hours of physical culture could not induce. The favorite mode of conveyance is the "isiwashchick," or four-wheeled vehicle closely resembling a victoria, with the exception that one horse is hitched between the shafts and another to the side of the first horse, but outside the shaft. The Cossack driver, perched unconcernedly on the seat, in green coat and puffed red sleeves, forms a ludicrous picture to the non-resident. The traveler, anticipating a delightful drive, takes his seat in the vehicle and almost immediately the horses, at a word from the driver, with a leap are off. The Cossacks are very proficient in driving and handling these conveyances. They start the animals very quickly and stop with a suddenness anything but pleasant to the passenger. They are dangerously reckless in driving and one must be gifted with considerable temerity to undertake a drive, especially on the roads of Vladivostok, where the prevailing conditions make driving rather more of an experience than a pleasure.

Russian Officers Ubiquitous.

In every hotel buffet, corridor and balcony one is confronted by the Russian officer; well built, natty, dignified and doubly proud and conscious of the fact that he is a servant of his lord, the Czar. The word "czar" is seldom used by Russians; they speak of their monarch as "Emperor." The Russian officer comes almost invariably from the well-to-do or aristocratic families and supports himself from his private income. He could not hold his social position otherwise, as the Czar does not pay his officers any too well.

Just the reverse to the officer is the Russian soldier—an inherent disciple of filth, dirt and squalor. He is irregular of feature and big of physique, but his stupidity is most pronounced and strongly verifies the fact of governmental neglect of the lower classes in the matter of education. But stupidity, however, has its moments of relaxation, even in a Russian soldier, brutish and ignorant as he is. The following incident, which happened in Vladivostok, will tend to illustrate the feelings of animosity against the Japanese which at present prevail throughout Siberia and Manchuria. There are numbers of Japanese in Vladivostok following various vocations, and short as they are they are not too small in stature as to be seen and ferreted out by Russian soldiers, who take fiendish delight in handling them roughly and intimidating them. One afternoon a short time before the war a big, burly Russian trooper, coarse and bloated in face and figure and a beard that none but a Russian dare exhibit, strutted up to a little, almond-eyed brown man on one of the main roads and demanded of him his passport. The Jap, not being gifted in the language of the country, explained in his best English, accompanied by a doff of his cap and a sweeping bow, that he had left his identification papers at home, whereupon the Russian, probably becoming incensed over the politeness of the Japanese, which possibly looked to him as affected, seized the little subject of the Mikado about the neck and violently shook him until his head actually rattled and marched him off to headquarters, in the meantime muttering between his teeth in the most broken English imaginable: "By-and-by fight, eh?"

Watching the Japanese.

Since the advent of the present Russo-Japanese imbroglio the Russian officials in Vladivostok and elsewhere in the orient have spared no pains and trouble in keeping a close watch on the Japanese in the country. It was well known to the officials that there were many Japanese within their city and thereabouts acting in the capacity of merchants, but who are in reality government spies. Because of these stringent measures adopted to keep the wily Jap from seeing too much there were frequent clashes between both Russians and Japanese. Imprisoned Japs oftentimes had recourse to the Mikado's diplomatic representative, which resulted in many cases in the alleged offenders being sent out of the country.

The Japanese found it exceedingly difficult to move about in any part of Siberia or Manchuria without the Russian officials becoming cognizant of the fact. This was the main reason why the Japanese have been supposed to be somewhat at sea concerning the quota of Russian forces in the far east at the outbreak of the war.

If the Russians of Vladivostok were ferreting out the Japs in Siberia and Manchuria they also lost no time in studying the strategical zones of the Japanese empire through the medium of government agents, who were either Germans, French or Englishmen. The most important fortified positions in Japan are the entrance to Nagasaki harbor, at Shimonoseki, the southern entrance to the inland sea; the extensive positions on the island of Awaji, which command the northern entrance to the inland sea, and the line of works commanding the entrance to Yokohama and Tokio. These positions were of vital importance to the Japanese, and they allowed no camera or sketching within a radius of several miles of any battery. During the first few months of the war several foreigners who happened to be in the vicinity of certain of these positions were taken into custody and had considerable difficulty in securing their release. The Japanese alarmist press spread the report throughout the country that several Russian spies had been detected in making maps of fortifications and had been arrested. Not a little excitement was caused thereby. So suspicious did the Japs become that any foreigner who happened to be in a district little frequented by foreigners was shadowed and followed by soldiers and police in disguise until he made his way back to one of the open ports.

Upon the deck of a small steamer in the little land-locked bay of Nagasaki, previous to starting on the trip to Vladivostok, there came aboard an elderly man, slovenly in appearance and unkempt of face, who took passage in the steerage. When the steamer arrived in Vladivostok this person was one of the first to disembark. The next afternoon the foreign visitors at Vladivostok were greeted with a most profound bow from a Russian colonel of infantry, and they recognized in him the indigent and impecunious individual who staggered aboard the ship at Nagasaki—a spy.

Vladivostok has been spoken of as a booming city. And booming it was—in a Russian sense. The Siberian railroad work calls for many supplies in the form of tools, machinery and provisions that are brought to and sold through Vladivostok, and from there are also sent goods and provisions used in a large back country. The demand for machinery and miscellaneous supplies was steadily increasing. Say all there is to be said of Siberia's natural resources—and it seems impossible to say too much—the industrial conditions are against the country becoming self-sustaining. The Siberian of the peasant and other common classes is too often lazy and shiftless, has no thought of the morrow and does not take kindly to farming life. Farming in eastern Siberia, which means Vladivostok and vicinity, under the most favorable Russian conditions will not curtail the demand for foodstuffs from the United States, for the reason that the population of that portion of Siberia is growing faster than the agricultural output.

The talk of wheat from the Vladivostok country competing with Pacific coast wheat is bosh. The kernel of the Siberian wheat is small and damp and makes poor flour, and the wheat cannot be improved. The finest samples of American wheat sown in this soil at once degenerates into Siberian wheat and Siberian wheat sown in California immediately yields the standard article of the golden state. Siberia is sui generis. The arctic plants and animals are seen in lower latitudes in Siberia than in any other country in the world and animals and plants peculiar to the temperate and even the tropical zone are found in southern Siberia. The tiger grows larger in Siberia and has a richer fur than his famous mate in India, and so in plant life.

No Place for a Foreigner.

As Siberia gains in population she will draw more and more heavily upon the outside world for machinery, manufactured articles and foodstuffs. A good country for the roving American? Not by any means. The foreigner who succeeds in Vladivostok must have rare tact, a good financial backing and the faculty of minding his own business. You enter the country on a passport, live in the country by official permission of the government and the police authorities permit you to leave. You cannot enter Vladivostok until the chief of police has seen your passport, and no agent will sell you a ticket for an outgoing ship or steamer until the chief of police has said you may leave. You cannot live at a hotel without registering your name, occupation, etc., on a police certificate, and you may be sure the landlord will promptly poke the paper under your nose and ask your signature.

Americans have succeeded and are succeeding in Siberia in a few instances, but their success is due to partnership with Russian officials and financiers, through whom business and concessions have been secured. No foreigner may buy real estate in Vladivostok. The law forbids it. A few pieces of land that were acquired years ago are held by Americans, and the smart advance in city lots in the last two years has inured to their advantage, but in future the Russian will be the sole beneficiary of his boom. Foreigners have to learn how to do business in Siberia, and the number of foreign houses is small. First of all you must respect Russian custom. You take off your hat when you enter a store and remove your overcoat and overshoes before entering house or office. Religion and loyalty are synonymous terms with the Russian. If above the common grade he is a stickler for formality. Society and government are both bureaucratic. No matter what hour of the day you make an official call or visit a high official you must wear your dress suit. This rule is imperative. The official hour for calling is 9 o'clock in the morning.

How the City is Ruled.

Vladivostok is nominally ruled by a mayor and council, who are elected by the few hundred residents entitled to vote, but the city is really dominated by a major general, governor of the maritime province of Vladivostok, a soldier of fine record and distinguished appearance, whose courtesy all visitors have cause to hold in grateful recollection. The day after he assumed charge of the district he was kind enough to spare half an hour from his official duties to talk of Siberia and its prospects. Like all other high officials in Siberia, he is ambitious for the speedy development of the country and believes that development will aid the commercial interests of the United States.

A week in Vladivostok gives a fair chance to catch the spirit of the place; to see that a large business calls for a rare combination of cleverness and commercial courage; to see that the merchant or mercantile agent must be equal to dealing in every commodity, from an eyelet to a machine for unloading coal at the rate of 100 tons an hour; to see with pleasure and pride two American sailing ships enter the harbor the same day and note that a single Japanese flag was the only other foreign signal in port.

As soon as the Siberian railway was finished and the fare from St. Petersburg to this point was fixed at the rate already decided upon—120 rubles ($60)—Vladivostok took on metropolitan airs. In an educational way the city makes an excellent showing. Two thousand children attend the seven schools. The general stores show the spring goods of late designs, quite up to our standard, and the two leading stores had ready-made suit, wrap and millinery departments where gowns and hats of modern style could be bought. The wives of the officers and merchants were more stylishly dressed than the majority of women seen in Shanghai, the swell city of the orient. The sailor hat has reached Siberia and is worn by women and children of all classes and all ages. Vladivostok matrons and maids are frequently dissatisfied with the designs of swagger American hatters, and too often mar the intended effect of the trim sailor by loading it with showy feathers and flowers. The Tam O'Shanter is popular with small girls and is made in colors that set the teeth on edge. An occasional tam band on the hat of a school girl bears the name of a Russian ship. The schoolboy, no matter how small, wears the cap and high boots of the accepted Russian dress, and his gray blouse is always confined with a shining black belt and metal buckle. His pranks on the street prove that he is a good running mate for his lively American brother.

No Exiles There.

One rarely meets any exiles as such in Vladivostok, but a daily view of as fine an assortment of jailbirds can be had as any place can show. The city jail is a one-story wooden building and stands in a yard bounded by a spike-topped wooden fence. The soldiers are on guard day and night and from hotel windows one could see sentinels marching to and fro on the other side of the street. On pleasant days the forty prisoners were escorted to church or taken for a walk, the desperate men in leg chains. And with due deference to the tales of Siberian prison horrors it seems that, as compared with the men closely confined in American prisons and jails, denied pure air and a chance to exercise, Siberian lawbreakers were picnickers.

A striking feature of the city landscape, on the bluff overlooking the harbor, is the granite shaft erected two years ago in honor of Admiral Nivelskoy, whose name is linked with the history of the Amoor River. Tablets recite the service rendered Russia by the brave sailor, and an eagle standing on the globe, his talons over the Russian empire, symbolizes the expansion policy of the Muscovite. The bust of the admiral sits in a niche facing the water.

When the American Asiatic fleet visits Vladivostok in the summer two of the ships go into the harbor at a time. American officers are always most cordially treated by the Russian officers there on duty.

It is exasperating to find yourself in a modern and bustling city where hardly a soul can speak a word of English or French or German or Japanese. One after another these languages may be tried with the coachmen and the policemen of Vladivostok, only to cause the stolid Russian faces to look at one more stolidly. Meanwhile the sun shines, the droskies tear by, the army officers in them salute, the tall horses blink at each other, the belted coachmen, with scarlet sleeves and women's dresses on them, make remarks with the cracks of their whips—everyone exchanges thoughts or words, while a European in the midst of this busy scene and in the midday sun is benighted—shut in from all the life about one as if by a high wall. It was a disagreeable and uncanny situation, and one from which no yellow-bearded coachman is concerned to relieve a traveler. If one is bold enough and addresses some of the tall military men, they, too, would have answered in French. Later an acquaintance with them and a bold attack on the Russian language of the streets make one able to get an inkling of what the people here were thinking of. But this growing intelligence on one's part did not extend to the Chinese and Koreans who infest the water front of Vladivostok or to the queer Siberians from Saghalien—ex-convicts thrown on the mercies of the town. All these ragged and unkempt swarms remain riddles (very dirty riddles, too), so far as the traveler's power of communicating with them are concerned.

Subtracting the Chinese and Korean elements from the streets of Vladivostok, and also the large number of those white men with a look of the "submerged tenth" about them, the town remains in appearances a squalid imitation of St. Petersburg. There is dirt everywhere, and upon every one, too, except the army and navy officers in their uniforms, and the Russian ladies. A close scrutiny of the latter's sisters in the lower class reveals an indifference to untidiness which is rather appalling. As for the men one meets on the streets they appear to enjoy dirt. The ischvostchiks are caked with dirt, their large and furious yellow beards are full of it. Their tall horses are rusty with it. Their vehicles have never been washed of it. Yet there is some excuse for all this, for Vladivostok during half the year, or when it is not frozen up, is a lake, a Venice of mud. The Russians are too careless and too anxious to spend money on more showy things than street pavements, such as buildings, dockyards, tall horses, furs and uniforms, to pay any attention to the cleanliness of the city.

Their footgear deserves a chapter by itself. Tall boots are generally worn, and they use a variety of leather overshoes, some of them reaching half way up the calf, for wading across the streets. The big Russian military men have in reality slender and well-shaped feet, but no one would guess it in Vladivostok. The only exception to the universal unconcern as to size and ungainliness in footgear is the Russian lady, who takes a pretty pleasure in picking her way through this muddiest of towns with her feet clad in Parisian slippers.

Monochrome of Mud.

The colors that dot the street seem like a remonstrance to the general monochrome of mud which threatens to absorb their brilliancy. There is no brighter scarlet than the shirtwaists of the coachmen, or brighter magentas and pinks than the women of the poorer classes wear. The white dresses of the nurses, trimmed with gay embroideries, are as gay as butterflies. The streets are at all times dotted with the uniforms of the army and navy, including the picturesque costumes which belong to the Cossack cavalry and the Siberian troops. There is quick movement in the streets, which adds to the gayety. The army officers are generally driving, and always driving fast. Their lives are said to be of a like pace, and the familiar tales of extravagance, official corruption, debt and dissipation which are whispered in every Russian city are heard in this remote spot of the realm. The officers look just as they do on the Nevsky Prospect of St. Petersburg.

There is a fine carelessness in the manner of Russians of distinction, both men and women, and here it is on "the world's street." One wonders if it has its root in some Slavonic strain of melancholy or some Asiatic trait of mind which feeds on the idea of the lack of importance of all human life. To the Russian temperament the idea of suicide is not so abhorrent as to us, and it did not seem extraordinary in Vladivostok to hear every week or two of some army officer who had deliberately taken his own life, after having with equal deliberation enjoyed a career at gaming, dissipation and debt to the point of insolvency. But no one would call these people a melancholy people who could see them on a holiday, of which they have a great many. The drinking of vodka is prodigious, the hilarity general. The harsh word intoxication should then be translated into merriment. Even the stolid ischvostchiks will get down off their droskies and waltz with one another in the mud.

Clubs and Much Music.

There are theaters and clubs and much music in this small rude town. Like all military towns far from home, the people do what they can to make life the reverse of dull. Nothing is on a small scale, whether in amusement or in serious business, and in Russia this applies even to the appearance of things. The men, the women, the horses are big. Contrasted with Japan, the land of mincing steps and little people and of jinrikishas, it is startling to be among this tall and swift crowd, where the stride of the horses as they tore up and down hill was in strong contrast to the ambling of the Japanese coolies.

The knowledge of the vastness of his empire makes the Russian accept with patience the long and weary toil of travel through it. The wonder is that in that little city, so many thousand miles from the great capital, he should appear so much a Russian as he does. Until within a few months he has been accustomed to get a tarantass, often with his family, crack his whip and start to drive 10,000 miles across Siberia. No wonder there is a patient look in his eyes. In this remote out-of-the-world spot, where one is yet not out of the world, but very much in it, no difficulty is experienced in reading upon this man's face his consciousness of the great Russia beyond, the Russia onward, far and very far away—Russia vast, Russia impressive, Russia restless, Russia benighted, Russia the mightiest of the mighty.

COMING OUT OF VLADIVOSTOK HARBOR.

The first effect of the Japanese attack upon the Port Arthur fleet was an order from Viceroy Alexieff calling out the Russian fleet at Vladivostok. There was a prompt response, the huge battleships rushing out of the harbor to harass Japanese shipping, threaten her coasts, and divide the Japanese fleet, which was blocking the harbor of Port Arthur.

ENGLISH CREWS GOING ABOARD JAPANESE SHIPS AT GENOA, ITALY.

Before the war the new Argentine cruisers, Moreno and Rivadavia, were bought by Japan. The scene represents the embarkation of a British crew on board the Nisshin, as one of them was renamed prior to its departure for Japan on the morning of January 9th, just a month before the engagement at Port Arthur.

CHAPTER XIX

JAPAN'S ARMY INVADES KOREA

Transports Loaded With Mikado's Troops Rushed to Gen-san, Masampho and Chemulpo and Take Possession of the Hermit Kingdom—Two Lines of Battle Formed and an Advance Made Toward the Yalu River on the North of Which Russia Was Concentrating a Large Army—Hardships of the Russian Troops Transported in Box Cars.

THE initial movement of the Japanese army was upon Korea, the prize of the war. The movement was attended by deep secrecy such as marked all the operations of the Mikado's fighting men on land and sea. Cables were cut and every means was taken to prevent the outside world from knowing what the Japanese were doing.

It was believed that the Russian troops in Manchuria would be pressed eastward as fast as possible to seize the Hermit Kingdom of Korea, and that the "Land of the Morning Calm," as it is called, would be the scene of the first great land battle. Transport after transport loaded with Japanese soldiers were rushed across the Korean straits under convoy of warships to Masampho on the south coast and Gen-san on the east coast of Korea, while others were sent to Chemulpo on the west coast, near Seoul, the Korean capital. The seizure of Masampho by the Japanese, which place they began immediately to fortify, was most important from a strategic standpoint. With Masampho and Fusan on the east coast of Korea heavily fortified, together with Japan's fortifications on her own soil on the opposite side of the straits, the connecting waterway between the Japan and the Yellow seas became a veritable Dardanelles.

Following the destruction of the Variag and Korietz at Chemulpo, the Japanese transports landed 19,000 Japanese troops at that point.

The Emperor of Korea, whose allegiance had been vacillating between the two contending powers, issued a manifesto granting the Japanese the right to invade his soil and was promised protection in return. He was also induced to declare the port of Wiju at the mouth of the Yalu river open to commerce. Before hostilities, Russia had protested against the making of Wiju an open port.

Arrangements were made to send a prince of the imperial Japanese house to Korea to act as Japan's viceroy, his nominal duty being tó advise the Seoul government.

By February 16, Japan had landed 120,000 troops in Korea. Of these 80,000 were extended along the fighting front south of the Yalu river.

This was their fighting front line which ranged from the great wall of China to Vladivostok. There were two lines, however. One reached from Chong-Yu to Kilju, from side to side of Korea, and was the strategic line south of the Yalu river. No fewer than 8,000 of Japan's finest fighting men were massed on this first line. Besides this first line there was a second, which ranged to North Seoul, with that city as its headquarters. Twenty thousand men were sent to this second line.

The Russians, on the other hand, showed no disposition to get far from their base of supplies on the Yalu, and prepared to intrench against a Japanese attack rather than to make a forward movement, on account of the insecurity of the coast line from the mouth of the Yalu river to Talien-Wan bay. Cavalry scouts were thrown out, however, and several skirmishes between the scouts and the invading Japanese marked the beginning of the conflict on land.

While the fighting lines in Korea were being advanced toward Manchuria, Japanese engineers disguised as Chinese laborers blew up a bridge on the Chinese Eastern railway, which interrupted railway communication with Port Arthur.

In the meantime there were reported attempts on the part of the Japanese to make landings for the purpose of cutting off the garrison at Port Arthur, but none of them was effective, if indeed they actually took place.

The Russians selected Harbin as the main base of the Russian land operations and troops were concentrated there as rapidly as the railway could transport them. The concentration of troops proceeded systematically and provision was made for the speedy arrival of 120,000 men from the divisions of Moscow, Kieff and Warsaw. The plans of Russia were made upon a large scale, the expectation being that she could put 100,000 additional troops into Manchuria in a fortnight. The total military strength in Manchuria as contemplated by the St. Petersburg authorities was 400,000 men. There were not more than 125,000 Russian troops in Manchuria when the war began. Every bit of available rolling stock was pressed into service and men were rushed to the front in coaches, box cars and freight trucks of every description. The Russian troops suffered many privations on account of inadequate transportation facilities, and from the intense cold which prevailed. Among the troops were several corps of Siberian rifles, crack Cossack regiments and a brigade of Cossack artillery armed with excellent mountain guns.

Instead of sending more troops to Port Arthur they were concentrated at Harbin, and as fast as they arrived from Russia they were dispatched to such points as required their presence. Some were sent south to New-Chwang, Antung and the points along the Yalu and others were sent east to Vladivostok, which the Japanese were expected to attack as soon as the ice went out.

Russia issued a formal protest to the powers against the Japanese invasion of Korea and expecially against the attack upon her two warships in the neutral harbor of Chemulpo. To this Japan replied at length, and as the reply covers all the reasons for Japanese occupation of the Hermit Kingdom and goes to the very heart of the controversy which culminated in the war, it is given herewith:

The imperial Japanese government is given to understand that the Russian government has recently addressed a note to the powers in which the government of Japan is charged with having committed certain acts in Korea which are considered by Russia to be in violation of international law, and all future orders and declarations by the Korean government are declared on that account to be invalid.

The imperial government does not find it necessary in the present instance to concern itself in any way with the views, opinions or declarations of the imperial Russian government, but it believes it to be its right and duty to correct misstatements of fact which if permited to remain uncontradicted might give rise to incorrect inferences and conclusions on the part of neutral powers.

Explains Landing in Korea.

Accordingly the government of Japan makes the following statement respecting the five acts which in the note referred to are declared to be fully proved and confirmed:

1. It is charged under this heading that "before the opening of hostilities against Russia Japan landed troops in the independent empire of Korea, which had declared neutrality."

The imperial government admits that Japanese troops landed in Korea before the declaration of war was issued, but not before a state of war actually existed between Japan and Russia. The maintenance of the independence and territorial integrity of Korea is one of the objects of the war and the dispatch of troops to the menaced territory was a matter of right and necessity, which had the distinct consent of the Korean government. The imperial government draws a sharp distinction between the landing of Japanese troops in Korea under the actual circumstances of the case and the sending of a large body of Russian troops to Manchuria without the consent of China, as was done by Russia, while peaceful negotiations were still in progress.

2. Under this heading it is alleged that Japan "with a division of the Japanese fleet, made a sudden attack on the 8th instant —that is, three days before the declaration of war—on two Rus-

sian warships which were in the neutral port of Chemulpo and whose commanders had not been notified of the rupture of relations, as the Japanese maliciously stopped the delivery of Russian telegrams by the Danish cable and destroyed the Korean government's telegraphic communication."

The imperial government declare that the allegations under this heading are untrue. The imperial government did not stop the delivery of Russian telegrams by the Danish cable, neither did they destroy the Korean government's telegraphic communication. Regarding the alleged sudden attack on February 8 on two Russian men-of-war in the port of Chemulpo it is only necessary to say that a state of war existed, and that Korea, having given her consent to the landing of Japanese troops at Chemulpo, the harbor of Chemulpo thereby ceased to be neutral, at least between the belligerents.

3. It is charged under this heading that "in spite of existing international laws shortly before the opening of hostilities Japan captured as prizes of war the Russian merchantmen which were in neutral ports of Korea."

The imperial government have established a prize court with full authority to pronounce finally on the question of the legality of the seizures of merchant vessels. Accordingly, it would be manifestly out of place for the imperial government to make any statement regarding the assertion under this heading.

Emperor of Korea Not Threatened.

4. It is asserted under this number that Japan "declared to the Emperor of Korea, through the Japanese minister at Seoul, that Korea would be henceforth under Japanese administration, and warned his majesty that in the case of noncompliance Japese troops would occupy the palace."

The imperial government declare the charge under this number to be absolutely and wholly without foundation in fact.

5. Under this heading it is charged that the Japanese govern-

ment "forwarded a summons, through the French minister, to the Russian representative at the court of the Emporer of Korea to leave the country with the staff of the Russian legation and consulate."

The imperial government deny the accuracy of this statement. No demand, either direct or indirect, was addressed by the Japanese government asking the Russian minister to retire from Korea. The French charge d'affaires called on the Japanese minister at Seoul and informed him verbally, as he did afterward in writing, that it was the desire of the Russian minister to leave Korea, and asked the opinion of the Japanese minister with reference to the matter. The Japanese minister replied that if the Russian minister would withdraw in a peaceful manner, taking with him his staff and legation guard, he would be fully protected by Japanese troops. He did so withdraw of his own free will on February 12, and an escort of Japanese soldiers was furnished him as far as Chemulpo.

The point of concentration of the Japanese army in Korea was Ping Yang, about 100 miles north of Chemulpo, and the same distance west of Gen San, at both of which points large bodies of troops were landed. The ultimate destination and base of was Wiju, a Korean port at the mouth of the Yalu, declared an open treaty port at the instance of the United States after the war had been begun.

The march to Ping Yang was most difficult, over roads which were a mass of slush in the daytime and frozen at night. As the Japanese army advanced it encountered thousands of Korean refugees fleeing southward from the anticipated Russian advance. Women with babies on their backs and men carrying household furniture trudged along in deadly fear, while the sturdy little Japs marched resolutely forward in an opposite direction to wage the same aggressive warfare on land they had already begun on the water.

CHAPTER XX

LOCKED IN THE BLACK SEA

Russia's Fine Squadron Barred From the Scene of War by the Treaty Governing the Dardanelles—Description of the Historic Strait Which Is the Key to the Turkish Capital and Connects the Sea of Marmora With the Aegean Sea—Fortified by the Ottoman Government and Considered Impassable Except With the Sultan's Consent.

THE historic Dardanelles early figured in Russia's naval operations. At the outbreak of hostilities Russia had a splendid squadron of warships in the Black Sea, consisting of five battleships, two cruisers and three torpedo boats. The battleships were the Restislav, 8,800 tons displacement, 16 knots speed, which has been increased to 18 by the use of petroleum on coal. It is armed with four 10-inch guns, eight 6-inch quick-firing guns, turrets, four torpedo tubes above the water. Tri Sviatitelia, 12,480 tons, 17.7 knots speed, four 12-inch guns, six 8-inch guns on broadside, smaller quick-firing machine guns in proportion, six torpedo tubes above the water. Dvanadsat Apostolof, 8,500 tons, 16.6 knots, four 12-inch and four 6-inch, and smaller guns. Tchesme and Sinope, sister ships, 10,181 tons, 17.8 knots, six 12-inch and seven 6-inch, and smaller guns. Kniaz Potamuin Tavritchesky, 12,500 tons, 16 knots, four 12-inch and sixteen 6-inch guns, with the usual secondary battery.

The cruisers were the Pamyet of 2,996 tons, and the Merkurya of 2,996 tons.

The torpedo boats were the Socken, Kazarsky and Guden, each of 400 tons.

This squadron was commanded by Vice Admiral Kruger, a thoroughly efficient officer, who gained his rank as vice admiral a year previous. He commanded the fleet which was sent to Bulgaria when Russia demanded reparation for the murder of her consul there. He came to the United States in command of the Russian warship Rynda in 1893, and commanded the naval guard on the Danube during the Russo-Turkish war.

When the Russian fleet at Port Arthur was so badly crippled by Japanese torpedo boats and in the subsequent naval battle, it was designed to send the Black Sea squadron to its assistance, but the only egress was through the Dardanelles, and this historic strait was barred to warships.

The Dardanelles itself is a little over forty-three miles long and is from three to four miles wide, but at the narrowest part is not more than 1,000 yards across. This is at a point guarded by Chanak Kalesi Castle, where huge chains are used to bar the passage. The strait connects the Sea of Marmora with the Aegean, and is supposed to be the key of the Turkish capital. This question of the protection of Constantinople is the central fact kept in mind by all the treaties governing the Dardanelles. Turkey has for centuries contended that no war ships should pass through the strait without her permission. She had built the first defenses at its southern end against the war galleys of the Venetians, and these she closely guarded. England was the first power to recognize Turkish rights in this respect. This was in 1809, and the question was not raised again until 1833, when the celebrated Russo-Turkish treaty of Hunkiar Iskelasi, signed June 26, closed the Dardanelles to all foreign fleets except those of Russia. The objection of the powers, especially of France and England, was so great, however, that the treaty was abrogated. Matters thus remained until 1841, when Mehemet Ali was making trouble for Turkey, and France was aiding him.

To keep the French fleet from menacing Constantinople a compact, called the quadruple treaty, was signed at London. formally closing the Dardanelles to foreign war vessels. This convention was signed by England, Russia, Austria and Prussia, and probably saved Constantinople from invasion. A few years later, however, France discovered that the treaty was an excellent arrangement to keep Russian ships from menacing her southern shores at any time, and she also signed the compact. Then the Muscovite discovered that unwittingly he had signed a convention to keep his war craft immured in the Black Sea. The Crimean war of 1854-56 clinched the matter. Nicholas I. was crushed by a coalition of the powers and was compelled to sign the treaty of Paris, which, while neutralizing the Black Sea and throwing it open to commerce, further declared that the Dardanelles should remain closed to foreign men-of-war. This attitude was confirmed in 1871, and again in 1878, after Turkey had been defeated by Russia. The treaty of San Stefano, signed March 3, 1878, granted to the Russian conquerors many concessions, including the free passage of the Dardanelles.

The powers objected to the arrangements, fearing that Russia had acquired a preponderance of power over the northern Turkish provinces, which might ultimately threaten Constantinople and the approaches to the Black Sea. After a critical period of diplomacy the congress of Berlin was held—June 13 to July 13, 1878—the result being a treaty that stripped Russia of much of the fruits of her victories over the Turks, and further declaring that the Dardanelles should remain closed to war vessels.

This treaty was signed by England, France, Prussia, Austria, Russia and Sardinia. Thus the matter has remained up to the present, except that in 1891 the porte and Russia reached an agreement that ships of the "volunteer fleet" of Russia bearing the flag of the merchant marine shall have free passage of the Dardanelles, but that Turkey must be notified should these vessels carry convicts or soldiers.

In January, 1903, Great Britain protested when four Russian torpedo boats from Cronstadt were permitted to pass the strait to join the Black Sea fleet, but none of the other powers took notice of the affair. It was obvious that when Britain's ally, Japan, was involved, England's protest against permitting Muscovite men-of-war to get out of the Black Sea, was of a more strenuous nature.

Soon after hostilities began the Russian government made overtures to the Sultan of Turkey for the release of the powerful Black Sea squadron despite the treaty of Berlin. It was reported and very generally believed that the Sultan was agreeable to a plan allowing the fleet to escape under the guise of merchant marine if he were given in return the privilege of carrying out his own policies in the Balkan States. The Turkish policy in the Balkans has been the persecution and massacre of Christians, which have met with opposition from Russia in every instance. To the credit of the Czar be it said that he is reported to have rejected every counter proposal of the Turkish government which would in any way enlarge its power to deal with the Christians in the Balkan States.

Meanwhile Russia was preparing her strong fleet in the Baltic sea for future operations in the East and rushing work upon vessels in every shipyard both at Baltic and Black Sea yards.

If it had been possible to release the Black Sea fleet immediately after the first attack upon Port Arthur there might have been a different story to tell of that historic harbor. As it was it is difficult to see how Russia's naval forces could have been worse disposed than they were. A strong squadron of her far Eastern fleet was ice bound in the harbor of Vladivostok, a port so strongly fortified that it was regarded as practically impregnable against a naval attack. One of her fastest cruisers, the Variag, had been detached from the main squadron and, with a little gunboat for a consort, was in the harbor of Chemulpo unable to defend herself against a Japanese squadron. Her most powerful squadron was at its home base in the Baltic, while many of her best ships could not be called into service by reason of the treaty which locked them in the Black Sea.

CHAPTER XXI

JAPANESE WOMEN AS WAR HEROINES

They Possess Courage Equal to That of the Spartan Wives and Mothers—Tender and Loving and Make Good Nurses in the Hospitals and on the Battlefields—Devoted to the Mikado and Deem it Honor That Their Sons and Husbands Should be Killed in His Cause—Anecdotes of Former Wars Which Display Remarkable Heroism and Devotion.

THE women of Japan in war time are not less heroic than the women of other nations, and there are many instances of their courage which equal that of the Spartan wives and mothers.

In the feudal times, which came to an end in Japan only thirty years ago, all gentlewomen were trained in the use of the sword and lance. The women of the samurai class received a regular military education and if the castle of a daimio was besieged, they were capable of assisting in the defense if necessary.

A noted instance of the martial prowess of the Japanese women occurred during the siege of the castle of Wakamatsu in 1869, where the Shogun made his final stand against the forces of the Mikado. Nearly one thousand women and girls belonging to the families of samurai attached to the Shogun fought behind the barricades and on the castle walls. Many of them were killed in battle, while not a few committed suicide rather than undergo the humiliation of defeat.

Yet the Amazonian qualities of the women of old Japan did not detract from their womanliness. They were tender mothers and loving wives. The nursing of the wounded and sick was part of the education of every samurai woman.

With the passing away of the age of chivalry in Japan, upon the downfall of the Shogunate, the Japanese woman was called upon to face new conditions, and how she met these conditions is shown in the history of the Chinese war of 1895.

It is a matter of record that some 10,000 Japanese women volunteered to go to the front as nurses in the field hospitals at the outbreak of the Chinese war, and advices from Japan state that the number of women volunteering to go to the front as nurses to-day is greater than in 1895. But the women who stay at home are not lacking in patriotic devotion.

There is an anecdote concerning the mother of the heroic Commander Sakamoto, who was killed on the bridge of his ship, the Akagi, at the battle of Yalu, which shows how the spirit of patriotism flames in the hearts of Japanese women.

An official of the navy department called on the family of the naval officer to convey, as delicately as possible, the news of his death. Having communicated his tidings to a member of the family, he was about to depart, when the shoji slid open softly and the aged mother of the dead commander staggered into the room.

She had been an accidental eavesdropper and had heard all. Trembling with emotion she bowed low to the visiting officer and said:

"Tell the Emperor I rejoice that a son of mine has been able to be of some service to him."

Some Japanese women refused to weep over their dead, because it was considered disloyal to the Mikado to weep for those who had had the honor to die fighting for him. When a wife or a mother heard that a husband or a son had been killed in battle, the first expression uttered, was an acknowledgment of the honor conferred upon her by the gods in being bereaved for the cause of the Emperor.

To the western mind such patriotism appears to be fantastic and hard to understand. In the light of Japanese history it does not seem so strange.

The spirit of patriotism in the Japanese women of the present generation is the outgrowth of ages of feudalism. The loyalty and devotion which the women of past generations gave to their feudal family head are in the present generation given to the Mikado.

In time of war the Empress of Japan sets an example for all the women of the country by her activities in behalf of all those who are suffering or in distress. She may be seen frequently visiting the great military hospitals, accompanied by a party of court ladies and noblemen's wives.

Following the example of the Empress, all the great ladies of Tokio society do what they can to relieve the distress and suffering that inevitably follow war. There is no class of women that does not contribute something to this cause; even those butterflies, the geishas, and the unhappy creatures in the Yoshiwaras give their share.

It is not only the women of the samurai class who show passionate patriotism in wartime. All classes of society are represented in the modern Japanese army, and the peasant woman has given proof that she is quite as devoted to the Mikado as the samurai lady.

A story is told of an old peasant woman who sent her only son to fight for the Emperor in the Chinese war. By depriving herself of everything but the barest necessaries of life, and toiling early and late in the fields, she had been able to give her son a superior education, and she had the satisfaction of seeing him fairly started on a business career, which promised to be successful. when the call to arms sounded.

The little peasant mother bade her son give up his business and enter the ranks of the army. The boy did as his mother wished, and his regiment was one of the first to set foot on Chinese soil.

Every morning just before daybreak the little peasant woman rose and, after making a careful toilet, as an orthodox Buddhist she went to a little shrine near by and prayed to Ojin, the god of war. She did not pray for her son to come home safe and sound, but she prayed that he might prove worthy of the honor of wearing the Mikado's uniform.

One day, when the old woman was returning to her home from the temple, she met a messenger who told her that her son had been killed in the attack on Port Arthur. The mother's eyes grew dim with tears, and she swayed unsteadily for a moment. Then she turned and started to go back to the temple.

"Where are you going?" cried the messenger. "Don't you understand what I say? Your son has been killed."

"Yes, I understand," said the old woman, calmly, "and I am going to thank Ojin for the honor he has conferred upon me."

The Japanese woman was busy from the beginning of the war. Hers is a patriotism that burns for active service. Always the Red Cross society of Japan is thoroughly organized and ready for service. With the actual breaking out of war its membership was increased by hundreds of thousands of women, who give not only their money but their time. None was too old, none too young to do something. In the headquarters that were established in every village it was a common thing to see women of 80 and children of 5 and 6 years old, all come to offer such service as they could give to the cause. Where it was necessary these women made personal sacrifice, gave up some luxury, some feminine fancy, to enable them to contribute more liberally. The empress herself set an example by working personally, preparing with her own hands bandages to be sent to the field hospitals and garments for the men at the front. At that season of the year the soldier's needs are much greater than in the months of milder weather, and there was work for all hands to do. The active personal interest of the empress is the same now as it was during the Chinese-Japanese campaign, at which time she not only worked while the war was in progress, but gave to each soldier who lost a limb an artificial one. In that war with China, as well as during the "boxer" troubles, it was demonstrated that Japan had a military hospital service unequaled by any nation in the world, with the possible exception of Germany, whose service Japan has taken as a model.

Japanese women of all classes employ professional hair dressers for the elaborate head dresses in vogue among them. Many of them, when the war broke out, learned to do their hair in foreign fashion, which they consider very ugly, in order that their contributions to the war fund might be the larger. Geisha girls sold their long silken obis, the most valuable articles of a Japanese woman's dress, for money to give to the cause.

War Widows of Japan.

One of the heroes who died for his country at Port Arthur was Lieutenant Miara, who volunteered to take in one of the transports that were intended for blocking the entrance. When he was killed nothing was left of him except his sword lying on the deck. He was awarded posthumous honors for his bravery.

His beautiful wife, to whom he had been married only ten days, shaved her head, as do all Japanese widows who do not intend to marry again, and entered a Buddhist nunnery.

There is a feeling among the Japanese that the widow of a man who has received posthumous honors is a kind of sacred person. But there is little doubt that these Japanese women love their husbands with a devotion and tenderness not excelled among western nations. Their Buddhistic faith is especially likely to encourage them to despise the idea of marrying again when they believe that after a long or shorter period they will be reunited to the object of their devotion for eternity.

The devotion of the Japanese women to the soldiers of the empire is illustrated by the following anecdote: A young soldier, awaiting marching orders, received a message from his mother the night before he left that she had selected a wife for him in the following manner: The girl had come with her mother, who said:

"Please take this, my humble daughter. Your son, I am told, is going to the battle. I congratulate you. To ask you to take this daughter so that she may take care of your house while your son is away is too forward. But this daughter wishes to nurse you in case you should become ill, that your son might go assured of your welfare."

Whereupon the soldier's mother had selected the girl as a bride for her son.

The patriotism of the Japanese women has been displayed in extreme cases by their work in loading and unloading cargoes and in donning men's attire to go and fight the foe.

In no country are children so much desired and treated with such affection—a gentleness that never spoils them. Sons are preferred, in the desire to perpetuate the family name; but the daughter is scarcely less welcome. Her prospective marriage, however, almost from the hour of her birth, is a source of anxiety. Upon that event she enters her husband's family, ceasing, practically, to belong to that of her parents, and her fate rests largely in the hands of her husband's mother, upon whom she must wait like a servant, assisting in the work of the house, sewing, preparing the food and serving the tea. No matter how unjust or unkind she may be, the son will never intercede for the wife with his mother, to whom he must render lifelong obedience. Should the wife be divorced, she loses her children, who remain with their father. There could be no other arrangement, since the mother could not support them, and would hardly impose the burden of their maintenance upon her own family. This law has induced the women of Japan, in common with those of other countries, to endure suffering, neglect and open infidelity, rather than be parted from the children.

In order to obviate the possible evils of an unhappy marriage, parents occasionally adopt a lad of intelligence and worth, to become at the proper time the husband of a tenderly loved daughter. In such a case the husband takes the family name of the wife, to whom the children belong, and his position is that of a subordinate and dependent. Such marriages are becoming less and less frequent, the modern, progressive Japanese objecting to the disability which a marriage of this sort necessitates.

RUSSIANS ENCAMPED IN A CHINESE TEMPLE.

IN THEIR military occupation of Manchuria during the war, the Russian troops respected nothing. They even broke into the Chinese temples and hustled their gods aside to make room for their arms and troops. But what could the native do except helplessly protest?

FRENCH INFLUENCE.

The early steps in the organization of the Japanese Army, on a modern basis, were conducted mainly by French officers. The drum-major therefore appears.

FURTHER GROWTH ILLUSTRATED.

In this picture the officer is seen instructing the sergeant major, while the orderly has doffed his cap according to modern practice.

CHAPTER XXII

FIRST ATTACK ON VLADIVOSTOK

Japanese Early in the War Make a Demonstration before Russia's Stronghold on the Japan Sea and Shell the Forts—Strong Russian Squadron of Four Armored Cruisers in the Harbor Fail to Reply—Japanese Land Troops Ninety Miles Below the City—Incidents of the Bombardment, which Was a Ruse of the Mikado's Squadron.

AT THE outbreak of the war Russia had a squadron of four powerful armored cruisers at Vladivostok, her stronghold on the Japan Sea, under command of Rear Admiral Stackleberg, an officer who won distinction in the war with Turkey, and who, before going to the far East, had command of the Imperial Yacht. Admiral Stackleberg's squadron consisted of the Rossia, Gromoboi, Rutik and the Bogatyr.

The Rossia displaces 12,300 tons, and its speed is 19.7 knots. On trial, it can steam 3,000 knots at full speed, while its armament consists of four eight-inch, sixteen six-inch quick-firing guns, twelve twelve-pounders, twenty three-pounders, and smaller guns in proportion, with five torpedo tubes and three third-class torpedo-boats carried on its deck.

The Gromoboi has 12,336 tons displacement and twenty knots speed, with practically the same armor as the Rossia.

The Rutik is a trifle smaller, with 10,940 tons displacement and a speed of 18.8 knots, and an armament consisting of a main battery of four eight-inch guns in sponsons on the upper deck and sixteen six-inch quick-firing guns on the main deck. These cruisers carried complements of more than 700 men each.

News of the attack on Port Arthur threw the city of Vladivostok into panic.

The people were terror stricken and all their thoughts were

bent on escaping. There were pitiable scenes at the railway stations, whence the authorities allowed 110 persons to leave daily by the empty troop trains that were going west. People fought and scrambled to reach the ticket office. Many who were unsuccessful knelt and prayed aloud, beating their breasts and offering any sum for a ticket. After the third day all traffic stopped and many started to walk inland.

Government Moved Inland.

The government offices, the bank, and the hospital were removed to Nikolisk. All shops were closed. Prices rose fabulously. The garrison, numbering 7,200 troops, was victualed for only four months.

The cruiser squadron left on Feb. 9 to attempt a dash around northern Japan to join the Port Arthur fleet, but a Japanese fleet was patrolling the seas between Saghalien and Tsugaru and the cruisers were forced to return. Their officers told of sinking a Japanese merchant steamer.

The latter part of the month the Japanese fleet landed a force of troops at Possiet Bay ninety miles below Vladivostok. The Russian fortress and the naval base of the Japanese action, in fact, would be exactly analogous to that of the Americans in the war with Spain when they occupied Guantanamo with an expeditionary force to serve as the base of their fleet blockading Santiago. This land force would be in a position to flank the main Russian army on the Yalu, threaten Vladivostok and at the same time to make a dash for Harbin if so ordered.

The landing of the troops at Possiet Bay was followed on March 6 by the appearance of the Japanese fleet off Vladivostok which began a bombardment of the fortress. The squadron consisted of seven warships which approached the port at ten o'clock in the morning and after discharging a few shots steamed out seaward.

The only building destroyed in this attack was an artisan's

cottage. A shell dropped through the roof, which fell. A wall was knocked down and fell, killing a woman, in the quarter of the town known as "Dirty Corner."

Another shell fell on the house of Colonel Skukeoff, without exploding until it passed through a bedroom, destroyed a stove, penetrated the wall, and passed into a room where there was a safe, with a sentry standing guard over it. Then it exploded, covering the soldier with earth.

He did not lose his presence of mind, but called out for some one to carry the regimental colors out of the house, which was done by the colonel's wife and the soldier.

In the courtyard of the Siberian barracks a shell exploded, slightly wounding five sailors.

It is supposed the attacking squadron consisted of a battleship, four armored and two unprotected cruisers.

The Russian batteries did not reply to the Japanese bombardment, which was due partly to the slight chance of hitting the enemy's warships at such a long range and partly to a desire not to betray the position of the batteries.

For fifty-five minutes in the afternoon Vladivostok was subjected to a terrific bombardment, in which four Russian sailors and the wife of an engineer were killed.

At 8:50 in the morning a sentry on the ramparts descried a thin cloud of smoke on the horizon beyond Askold island, at the eastern entrance of Usuri bay, about thirty-two miles southeast of Vladivostok. A few moments later it became apparent that the smoke was from a fleet of approaching vessels.

Scores of officers with glasses in their hands thronged the walls of the fortifications and eagerly studied the oncoming fleet. It was more than an hour before it could be clearly made out. As it came nearer it was seen that the fleet consisted of seven Japanese warships, four battleships and two cruisers, accompanied by two torpedo-boats.

When it became evident that the approaching vessels were warships the entire garrison was called to arms and there was a scene of great excitement. In a few moments every battery was manned.

General Veronitz and General Artamonoff made every preparation to repel the enemy and then awaited the attack.

Slowly the fleet steamed westward from the open sea until at noon it had reached a point midway between Askold island and the coast. The water of Usuri bay was covered with ice and this considerably impeded the advancing squadron.

As they neared the harbor the Japanese ships formed in line of battle and at 1:25 opened fire on the fortifications, steaming slowly along the water front of the city.

For hours the women and children and other noncombatants had been hurrying to the hills in the rear of the city, and when the first gun was fired there were few but soldiers within range.

For fifty-five minutes the bombardment was kept up, every ship in the Japanese fleet taking part. During all this time the Russian batteries were silent, reserving their fire for the nearer approach of the enemy.

The Japanese admiral, however, took care to keep his fleet out of reach of the shore batteries, and at no time did the attacking warships come within a mile and a third of the shore.

Shells Do Little Damage.

The sound of the cannonading was terrific, but the lyddite shells, of which fully 200 were thrown into the city and the fortifications, did comparatively little damage, as most of them failed to burst.

Finally, at 2:20, the Japanese fleet drew off, without a single gun having been fired by the Russian batteries. As the warships neared Askold island they were joined by the two torpedo-boats that had remained in the rear and simultaneously two others approached from the direction of Cape Maidel. Rounding the island the fleet disappeared from view, the attack having been practically without result.

Two of the Japanese vessels were the first-class cruisers Idsumo and Yakumo. The others could not be identified. All were covered with ice.

The attack cost the Japanese more than the Russians, for they used up at least $100,000 worth of ammunition without inflicting any serious damage either on the city or the fortifications.

The Japanese squadron reappeared before Vladivostok on the morning of March 7, but did not again attack the forts. These movements, as it was afterwards learned, were to cover the landing of more troops at Possiet bay and the seizure of Askold island as a naval base.

The official report of Viceroy Alexieff upon the Japanese attack is as follows:

"I have the honor to communicate to your majesty the following details of the events of March 6.

"The enemy's squadron approached Vladivostok toward 11 o'clock in the morning, having passed near Askold island. After several maneuvers which involved changes in the squadron's order of battle two cruisers were left to the north of the island and the remaining vessels of the squadron steamed along the coast of Usuri bay, parallel to the shore, keeping about fifteen versts (approximately ten miles) therefrom.

"Upon arriving off Mount St. Joseph and the Usuri bay battery the squadron, preserving the same order, made toward the battery. The ships opened fire from both sides, evidently first using blank cartridges in order to warm their guns. At 1:35 p. m., when at a distance of eight versts (approximately five and one-quarter miles) from the shore, the leading ship opened fire with her forward guns, and then the entire squadron steamed along the shore, firing their port guns as they went. The enemy did not fire while turning.

"After the third turn the squadron, at 2:25 p. m., ceased firing and steamed off to the southward, about ten miles to the right of Askold island, finally disappearing at 5:30 p. m.

"In all the enemy fired about 200 shells with no effect. No

damage was done to the fortress or the entrenchments, and in the town and at other parts of the fortifications the damage was insignificant.

"The garrison is in excellent spirits and the operations of preparing the batteries for action were carried out in perfect order.

"According to reports of the events of March 7 the enemy's squadron reappeared at 8 o'clock in the morning near Vladivostok. They entered Usuri bay and proceeded along the coast without opening fire. The squadron then returned and headed for Cape Gamova (Possiet bay), which it reached at 3:40 p. m.

"The enemy finally turned off when opposite Pallas bay and departed in a southerly direction. "Alexieff."

CHAPTER XXIII

STORY OF THE JAPANESE HOBSONS

Admiral Togo's Men Load Five Merchant Ships with Stone and Explosives, and Under Convoy of Torpedo Boats Attempt to Sink Them in the Channel at Port Arthur and Bottle Up the Russian Fleet—Affair Similar to Hobson's Exploit at Santiago, but Ends in Partial Failure—Thrilling Experiences of the Heroes of the Daring Deed.

THE most thrilling incident in the long naval siege of Port Arthur was made on February 24, when the Japanese attempted to duplicate Lieutenant Hobson's feat at Santiago during the Spanish-American war, by sinking merchant vessels in the narrow channel, thus bottling up the Russian fleet.

It was as bold and daring in execution as anything ever attempted in naval warfare, and while it was only partly successful, it will always stand out as a rare exhibition of strategy and bravery. The most remarkable fact in connection with the daring and hazardous feat was that not a Japanese life or vessel was lost, except the sunken merchant ships.

The Japanese loaded five steamers with stone and explosives and supporting them with the torpedo-boats and cruisers of the fleet, put on all steam and headed for the mouth of the harbor.

The watch on the disabled battleship Retvizan, lying nearest to the entrance of the harbor, was the first to discover the approach of the enemy. Though unable to put to sea, the battery of the big warship was intact, and in a moment her huge guns were playing furiously upon the approaching steamers, two of which seemed to be heading directly for her.

It was but a moment before the land batteries and the guns of every vessel in the Russian fleet were in action. The Japanese warships of Admiral Togo's fleet, following closer in the wake of the stone-laden steamers, were not slow in replying, and the darkness of the night was lighted with the flashes of the guns from the opposing fleets.

The daring Japanese sailors on board the steamers that were to be sacrificed in the attempt to block the harbor kept boldly on. In the darkness the Russians mistook the foremost vessels for the cruisers of the Japanese fleet and centered their fire upon them.

Sailors Take to the Boats.

Shot after shot pierced the wooden sides of the vessels, and they were rapidly filling with water before the crews attempted to escape.

Two of the vessels were sunk near the entrance to the harbor and a third went aground near the lighthouse on the little peninsula known as the Tiger's Tail.

Just before the vessels sunk the sailors were seen lowering the boats, in which they succeeded in reaching the warships of the fleet.

After keeping up the fire for some minutes longer the Japanese fleet drew off.

Japanese naval officers commanded and "jackies" manned the five steamers that were sunk at the entrance to Port Arthur. Merchant sailors volunteered for the dangerous undertaking, but their services were not accepted. Admiral Togo desired to intrust the perilous mission only to the navy.

The Japanese Hobsons.

The naval officers who commanded the five merchantmen were: were:

Commander Ryoketsu Arima.
Lieutenant Commander Taker Hirose.
Lieutenant Scichigoro Sailo.
Lieutenant Yoshita Masaki.
Sublieutenant Yasuzoto Tousaki.

Chief Engineers Daizo, Yamaga, Tomitaro, Kurita, Yasou and Minamisawa and Assistant Engineer Chikanon, Oshishi, Masando and Sugi handled the five engines.

The five merchant vessels which were prepared for sinking were the Jinsen Maru, 2,331 tons; the Tien Tsin Maru, 2,943 tons; the Hokoku Maru, 2,776 tons; the Buyo Maru, 1,609 tons, and the Bushik Maru, 1,399 tons. Each carried five men, two steering and three firing and running the engines.

There were ten officers and sixty-seven sailors in the crews, and all volunteered for service. They bade farewell to their comrades, expecting to die under the fire of the batteries of the enemy. The rescue of the entire crews surprised even the Japanese, who expected that a majority of the daring seamen would be killed.

The steamers did not carry lights and were not armed, and, consequently, were not discovered until the operation of sinking them was practically completed. Japan was loud in singing the praise of the volunteer crews who participated in the dangerous work.

Naval Officers Vary Report.

The official report of the naval officers who attempted to block the entrance to the harbor of Port Arthur the morning of February 24 differs in some respects from the other accounts. From this report it appears that the Russian searchlights discovered the approach of the five Japanese steamers before they had reached the point where it was proposed to sink them, and that the Russian guns disabled three of the five. Another feature was that a portion of the crews of the sunken steamers were not picked up until the afternoon of the 24th. The report of the officers in substance is as follows:

"Five vessels intended to obstruct the entrance to Port Arthur advanced about 4 o'clock on the morning of the 24th from the southward through the Lao-Thi-Shan channel toward the mouth of the port. It appears that the Tenshin Maru, which was in the van, was steered too far to port, and when she was about three miles to the southwest of shore she was shot and seriously damaged. She was run on the shoals intentionally.

"The other steamers, which were in her wake, changed their course to the northeast and advanced, but the enemy's searchlights, flashing upon them, impeded their progress. The enemy's fire struck the steering gear of the Bushiu Maru, disabling her. She grounded near the Tenshin Maru, and after striking, her officers destroyed her and she sank. Next the Buyo Maru was seriously damaged by the enemy's shells and she sank before reaching the harbor entrance.

Explosives to Destroy Ships.

"In the meanwhile, the Hokoku Maru and the Jinsen Maru had advanced with great rapidity and had reached the entrance to the harbor with considerable difficulty. The Hokoku Maru was on the outer side of the stranded Russian battleship Retvizan and the Jinsen Maru on the eastern side of the Retvizan. The crews of each lighted explosives to destroy the merchantment, and after giving a loud cheer, got into their small boats.

"When they found their vessels sinking the crews endeavored to row to the Japanese torpedo-boats, which were ready to pick them up, but the Russians' searchlights lit up their path and the Russian fire became very severe. The crews in their small boats were compelled to row around under cover, and they were unable to reach the torpedo-boats. The sea became heavier at sunrise and the crews suffered great hardship. They finally succeeded in reaching the Japanese squadron at 3 o'clock in the afternoon of the same day."

Russian Official Report.

The Russian version of the affair is as follows:

"At a quarter before 3 in the morning of February 24, numerous Japanese torpedo-boats attempted to attack the battleship Retvizan and sink large steamers loaded with inflammables.

"The Retvizan was the first to observe the torpedo-boats, and opened a strong fire on them. It was supported by the land batteries.

"It destroyed two steamers near the entrance of the harbor; they were coming directly toward it. One of them went on the rocks near the lighthouse on Tiger peninsula and the other sank under Golden Hill.

"The Retvizan observed four steamers in a sinking condition, and eight torpedo-boats departing slowly to rejoin the waiting Japanese warships.

"A portion of the crews of the Japanese vessels was drowned.

"The grounded steamer is still burning.

"The enemy is observed in the offing of Port Arthur in two lines.

"The Japanese crews saved themselves in boats, and it is possible that some of them were picked up by the enemy's torpedo-boats.

"I am proceeding to examine the coasts. The entrance to the harbor is open.

"I attribute the complete derangement of the enemy's plan to the brilliant action and destructive fire of the Retvizan.

"Floating mines are still visible in the roadstead. I have recalled the three cruisers sent in pursuit of the enemy, in order, in the first place, to clear the roadstead of the floating mines.

"We had no losses. "Alexieff."

It is evident from the foregoing that the Russians had no real conception of the Japanese plan to block the harbor entrance, and imagined that it was simply another early morning attack upon the stranded battleship Retvizan. Hence, the news was received in St. Petersburg as a great Russian victory, and the populace rejoiced in the belief that the Japanese squadron had been repulsed and that a number of the enemy's warships had been sunk.

Although every one of the Japanese Hobsons were rescued, some of them underwent extreme hardships before they succeeded in rejoining their mates, particularly the men on the Jinsen Maru, commanded by Lieutenant Saito, and the Bushin, commanded by Lieutenant Tarisaki. Lieutenant Saito said that his ship got so close to the Retvizan that had his men been armed with rifles they could have fired into the Russian crew.

When these crews found that the scheme was a failure, they dropped anchor and the men crowded into the remaining boats. They then blew up the steamers. A strong wind and the glare of the lights prevented the men from reaching the torpedo flotilla, and at daylight they were out of sight of the fleet, having been driven in an easterly direction by the wind to the Miaotao islands, which they reached the same evening.

The boats did not land together, but the men found each other on the islands. They chartered a junk, in which they went to Tungchowfou. They walked forty miles to Cheefoo, going two days without food.

After taking to their boats, they were for a long time under the fire of the Russian vessels and land batteries. Every few moments the glare of a searchlight fell uopn their little boats, making them a fair target for the enemy. Their position was much the same as that of Hobson and his men in the channel of Santiago. While both attempts were failures, they were attended with the same risks, although the Japanese had the better luck not to be captured.

The total number of men who took part in the attempted blockade on the merchant ships and torpedo-boats was 67, and of these 65 were promoted to be officers.

CHAPTER XXIV

THE SIEGE OF PORT ARTHUR

Admiral Togo's Repeated Attacks Upon the Russian Stronghold Keep the Russian Fleet and Forts Constantly on the Defensive—Many Futile Attempts to Block the Channel Result in Sharp Fights Between the Forts and the Torpedo Flotilla—A Desperate Conflict in the Open Sea Between Russian and Japanese Torpedo Boats.

THE siege of Port Arthur by the Japanese squadron under Vice Admiral Togo was the longest, most memorable and most incessant of all the naval conflicts of the war. Day after day Togo continued to shell the town and made many daring attempts to bottle up the Russian squadron by sending stone-laden fire ships into the narrow harbor neck.

Following his first attack he was aided in his work of destruction by the accidental loss of two Russian warships, one a torpedo transport and the other a second-class cruiser, which were blown up by submarine mines planted by the Russians themselves.

The first of these, the torpedo transport Yenisei, was laying mines on February 11 in the entrance to Talien-wan Bay to close it against attack. Her orders were to plant 400 mines and she had accomplished the task of planting 398 of them before the disaster which destroyed her and her crew occurred. The three hundred and ninety-ninth floated instead of sinking. On this the Yenisei drew off and fired at it with her light guns, attempting to sink or explode it. While thus occupied the lookout man discovered that the ship had drifted close to another mine. He gave the alarm and jumped overboard. Immediately there was a terrific explosion. One hundred and ninety officers and men, including the captain, were killed, while ninety more in the ship were picked up by boats and saved. The officer laying the mines, who was the only man in the Russian fleet with a knowledge of exactly how the mines were disposed, was among the killed.

A violent storm followed, when the mines came to the surface and floated about in all directions. The next day the Boyarin was sent to assist in securing them. She was caught in the storm, however, and was driven upon another mine and wrecked in the same manner as the Yenisei. This made a total loss of Russian vessels eleven in the first week of the war.

Viceroy Alexieff's official reports on the disaster said that Captain Stepanoff, the commander of the Yenisei, the engineer, two midshipmen, and ninety-two of the crew perished. The Russian second-class cruiser Boyarin, which was blown up by a mine February 12 in the same manner as was the Russian torpedo transport Yenisei, had on board 197 officers and men, all of whom were lost.

The Boyarin was 348 feet long, 41 feet beam, and 16 feet draft. It was 3,200 tons displacement, and its trial speed was twenty-five knots. The armament consisted of six 4.7 inch guns, eight 1.8 inch guns, two 1.4 inch guns, and three machine guns. The cruiser also was fitted with six torpedo tubes. The Boyarin was last reported as having taken part in the first battle at Port Arthur.

The Yenisei was specially designed for the work of laying submarine mines. Naval experts declared that the destruction of the Yenisei looked as if some one had been guilty of a gross act of carelessness. Such mines, it was pointed out, are controlled from shore and ought to be disconnected while a vessel is engaged in picking up a loose one.

The Yenisei was built in the Baltic works. It had a displacement of 2,500 tons and carried an armament of five twelve pounders and six three pounders, quick firing guns. The vessel was 300 feet long, 40 feet beam, and drew 14 feet 6 inches.

Commander Stepanoff, on seeing his ship, the Yenisei, in contact with his own mine, ordered the crew to the boats. The men refused to go, and Stepanoff then drew a revolver and threatened to shoot if they did not try to save themselves immediately. His last words were, "This is how one dies when he has luck." Stepanoff was a specialist in mines, and one of the most valuable officers in the Russian navy.

After the failure of Admiral Togo to block the harbor entrance in imitation of Lieutenant Hobson's feat at Santiago, he renewed his attack upon Port Arthur and succeeded in drawing the enemy's ships into a sharp battle, resulting in the loss of a Russian torpedo boat, but with few casualties. The nature of the fight is described in the official reports of the Russian and Japanese commanders.

Viceroy Alexieff reported as follows:

"After the moon had set early in the morning of February 25, the Retvizan repelled several attacks by the enemy's torpedo boats, two of which are believed to have sunk in the open sea.

"Our torpedo boats, under Captain of the First Rank Matussevitch, and Captain of the Second Rank Prince Lieven, unsupported, encountered and pursued the enemy's torpedo-boat flotilla. They sighted no large warships.

"Later in the morning of February 25 the cruisers Bayan, Diana, Askold and Novik were sent out to prevent the Japanese cruisers from pursuing a portion of our returning torpedo-boat flotilla. One of our torpedo boats, which was cut off by four Japanese cruisers, sought shelter in Dove (Pigeon) Bay, where it was subjected to a long distance fire by the enemy. It had no casualties.

"The Japanese fleet on sighting our cruisers came in closer to the forts, which, together with our warships, opened fire at 10:50. Our cruisers, still firing, entered the harbor, which our torpedo boats had already safely reached.

"The enemy's shells for the most part fell short. One seaman was wounded, but we sustained no other casualties.

"The Japanese fleet consisted of seventeen large warships and eight torpedo boats, whereas the squadron which attempted to block the entrance to Port Arthur on the previous day had twelve torpedo boats."

The Japanese report was written by Vice Admiral Kamimura, division commander under Admiral Togo. The report said:

"A bombardment began at long range, and at 11:45 a. m. all the ships and batteries were responding vigorously.

"Shortly after noon the Novik retreated into the inner harbor. The Askold and Bayan quickly following, demonstrated that the sinking of the steamers had not blocked the entrance of the harbor. A bombardment of the inner harbor was then ordered, and for fifteen minutes all the heavy guns of the Japanese fleet threw shells over the hills into the harbor.

"The Japanese were unable to determine the effect of the bombardment. but saw huge columns of smoke arising from time to time.

"In the meantime the Japanese cruiser squadron discovered two Russian torpedo-boat destroyers at the foot of Laotche Hill and gave chase. One of the destroyers escaped, but the other was pursued into Pigeon Bay, where it was sunk. The Japanese fleet sustained no damage and did not lose a single man."

Again on the last day of February the Japanese fleet of fifteen warships made another terrific attack and furiously bombarded the Russian stronghold and fleet for two hours.

The attack began at 10 o'clock in the morning. A few minutes before that hour the Japanese fleet of fifteen warships was seen steaming rapidly from the direction of Dalny. The Russian cruisers Askold, Novik and Bayan and four torpedo-boats were sent out to meet the enemy.

Fighting began at long range and after a few minutes of furious cannonading the Russian ships were forced to retire, the Novik and the Askold sustained serious injury.

It could be plainly seen from the land that the Askold and the Novik were both hit. One of the Russian torpedo-boats sank off Lighthouse Point.

The Japanese fleet pressed closely behind the fleeing Russians and as they reached the shelter of the rocky wall surrounding the

RUSSIAN OFFICERS HURRYING TO THE FRONT.

NEAR here is Lake Baikal, which was the scene of one of the great horrors of the war. Thousands of soldiers perished in the arctic storms which swept over its face, or were swallowed by gaping fissures while being hurried to the front.

TORPEDO ATTACK ON PORT ARTHUR.

IN THE above are illustrated: (I) the Japanese attack, indicated by arrows; (II) the torpedo net under water being penetrated by torpedo cutter; (III) the Brennan torpedo for harbor defense; (IV) the spar torpedo; (V) sections of the Whitehead torpedo, used by the Japanese: (VI) explosive head of the Whitehead; (VII) modern torpedo tube, in sections, used in Japanese Navy.

harbor turned its guns on the land batteries that crown the heights on each side of the narrow entrance.

For two hours the bombardment continued, shells from the enemy's ships falling fast in the beleaguered city. It was evident that the Japanese were attempting to throw their shells over the high wall of rocks surrounding the harbor and destroy the Russian fleet that had sought shelter there.

Finally, a few minutes after noon, the Japanese fleet drew off in good order, having sustained little damage that was apparent from the shore.

The superior range of the Japanese guns was again demonstrated, for while they were able to reach effectively the Russian ships the shots of the latter seemed to fall short of the mark. The same defect characterized the work of the Russian land batteries.

Much damage was done in the town by the shells from the Japanese warships.

Before the next important naval engagement connected with the siege of Port Arthur was fought Admiral Stark, who had commanded the Russian fleet, was supplanted by Vice Admiral Makaroff, whose daring exploits on fast warships had won for him the title of "the Cossack of the Sea."

Makaroff's Brilliant Exploit.

When the Japanese fleet of torpedo boats reappeared before Port Arthur on March 10, Admiral Makaroff sent out his torpedo boats to meet them and a severe clash followed, which is fully described in the official dispatches of Admiral Makaroff, transmitted to the Czar by Viceroy Alexieff. The first of these is as follows:

Mukden, March 11.—Admiral Makaroff, commanding the Russian fleet, reports from Port Arthur under date of March 11. as follows:

"Six torpedo-boats which went out to sea the night of March 10, four of them being under the general command of Captain Mattoussevitch, encountered the enemy's torpedo-boats, followed by cruisers.

"A hot action ensued, in which the torpedo-boat destroyer Vlastini discharged a Whitehead torpedo and sunk one of the enemy's torpedo-boats.

"On the way back the torpedo-boat destroyer Stereguschtchi, commanded by Lieutenant Sergueieff, sustained damages; her engine was disabled and she began to founder. By 8 o'clock in the morning five of our torpedo-boat destroyers had returned.

"When the critical position of the Stereguschtchi became evident I hoisted my flag on the cruiser Novik and went with the Novik and the cruiser Bayarin to the rescue. But as five of the enemy's cruisers surrounded our destroyer and as their battle-ship squadron was approaching, I did not succeed in saving the Stereguschtchi, which foundered. Part of the crew was made prisoner and part was drowned.

"On the ships which participated in the night attack one officer was seriously and three others were slightly wounded, two soldiers were killed and eighteen were wounded.

"At 9 o'clock this morning fourteen of the enemy's ships assembled before Port Arthur and a bombardment was begun with the heavy guns of their battle-ship squadron at long range.

"This lasted until 1 o'clock in the afternoon. It is estimated that the enemy fired 154 12-inch shells. The damage to our vessels was insignificant, and they are again ready for battle. Our losses were one officer slightly wounded and one soldier killed and four soldiers wounded.

"The illumination of the sea at night by the searchlights mounted at our batteries was most satisfactory, and several times isolated shots from our batteries forced the enemy's torpedo-boats to retire.

"With the commencement of the bombardment at dawn the guns of the fortress replied to the enemy's fire.

"The crews of all the ships engaged gave proof of remarkable coolness in action. Below decks the work of the day followed its ordinary course in spite of the shells falling between the vessels and covering them with fragments.

"A bombardment at such a distance must be considered ineffective, but the Japanese cruiser Takasago is reported to have been seen to suffer serious damage, the extent of which, however, it was impossible to ascertain at a distance of five miles. Many shells were fired at a range of seven and one-half miles.

"I have the honor to report the foregoing to your Majesty.

"Alexieff."

Russian Officers Wounded.

Viceroy Alexieff also sent the following message to the Emperor:

"In the fight between our torpedo-boats and the Japanese cruisers March 10 Captain Mattaussevitch, Ensign Alexandroff and Mechanical Engineer Blinoff received slight wounds and Ensign Zaeff was severely wounded in the head, losing his right eye.

"The commandant at Port Arthur reports the following details of the bombardment of the fortress there March 10.

"As soon as the enemy opened fire our batteries replied. Six of the enemy's ships remained behind the Leao-Tishin promontory and opened fire on the fortress over that shelter. They ceased bombarding at 1:15 p. m. The enemy fired about 200 projectiles. One shell from battery No. 15 on Electric cliff damaged a Japanese cruiser seriously.

"The results of the bombardment were insignificant; six soldiers were wounded. Three inhabitants of the town were killed and one was seriously wounded.

"According to General Stoessel's report the officers and soldiers in the shore batteries displayed exemplary courage and fired their guns in perfect order. Alexieff."

As far as is known this is the first time torpedo-boats have engaged each other at sea. Although the odds were against the Russians, as the Japanese flotilla was supported by the cruiser squadron, the Russians made a heroic dash for the foe and apparently had the better of the combat, sinking a Japanese torpedo-boat, until the cruisers got within range and one of the latter's shells crippled the Stereguschtchi.

The gallant action of Vice Admiral Makaroff in transferring his flag to the fast cruiser Novik and sailing out in the face of the enemy in an attempt at rescue received unstinted praise everywhere, stamping him at the outset of his command as a man of force and action, who insisted on being in the van of the fighting.

Vice Admiral Makaroff later visited the torpedo-boat destroyers which participated in the fight with the Japanese March 10 and warmly thanked the officers and crew for their splendid behavior. The admiral distributed decorations.

A correspondent of the London Times who was with the Japanese forces gives a more graphic account of the battle between the torpedo-boats than is contained in the official dispatches. According to his account, the vessels were so close together that the Russians threw charges of explosives onto the bridge of one of the Japanese torpedo-boat destroyers. These, however, failed to detonate. All the vessels engaged were more or less damaged.

The Japanese losses were six killed and eight wounded. That the Russians were defeated in spite of their superior numbers is due to the better shooting of the Japanese and the fact that the Russian vessels were armed with three-pound guns, while the armament of the Japanese ships was made up of six-pounders.

In reference to the fight of three hours which occurred later, the correspondent says the Russians fought with desperation and the Japanese with confidence born of their past victories. One Russian commander was killed early in the fight. A lieutenant then took command, only to fall, shot in both legs. Then the command devolved upon the sub-lieutenant, who also was killed after taking the wheel himself. When the coxswain fell this vessel was captured by the Japanese. The other Russian vessels escaped.

On the Japanese side one destroyer was hit on the water line, two of her compartments were flooded and her quick-firing ammunition was soaked with water. This vessel retired from the action. Her officers escaped narrowly from a twelve-pound shot which struck the platform in front of the bridge, killing one man and sweeping the bridge with splinters. The same vessel received a three-pound shot through the hull, but the damage was repaired within four days.

A Japanese lieutenant who boarded the Stereguschtchi said he had never seen a more sickening sight. Thirty bodies, terribly mutilated by a shell, were lying on deck. As the Japanese approached two Russian bluejackets rushed from the conning tower, locked themselves in the cabin aft and refused to come out. Two stokers jumped overboard and were picked up. These, with two wounded men, were the only survivors of a crew of fifty-five. When the Stereguschtchi sank the men who were locked in the cabin sank with her.

The fight lasted nearly an hour. The weakness of the Russian torpedo-boat destroyer armament again proved fatal. The Japanese concentrated their fire first on the Russians' twelve-pounder and put it out of action early in the fight, leaving her with only her three-pounders against their own twelve-pounders and six-pound guns. The Japanese had three killed and four wounded.

Describing the bombardment of Port Arthur the same day, the correspondent asserts that more than 110 shells fell in the town. The effect of the great shells from the twelve-inch guns must have been appalling. Outbreaks of fire were seen and the report of an explosion was heard, which it was surmised occurred in one of the magazines of the forts.

The first inside account of the damage done at Port Arthur by the bombardment which followed the sea fight was brought to Shanghai by three Norwegian steamers which left Port Arthur a few days after the battle.

These vessels were chartered by the Russian admiralty as coal transports. During the bombardment the Argo lay alongside the Retvizan in the inner harbor. A shell from a Japanese warship fell on the deck of the Retvizan and exploded, killing nineteen officers and men.

The crews of the merchant ships in the harbor deserted and fled towards the promontory during the bombardment.

The greatest damage was done to the new town, where the shells caused such havoc that practically all shops and business buildings in the main street were demolished. Scarcely a residence was left untouched.

Sight Seers Blown to Pieces.

The Japanese fire was marvelously accurate. The inhabitants were terror stricken, and many attempted to construct rude bomb-proof shelters.

One shell fell among a crowd of people who were gazing at the attacking fleet. It killed twenty-five. Three government clerks were killed while hurrying from the port admiral's office.

The most elevated fort at the entrance to the harbor was most seriously damaged.

A two-funneled cruiser that was anchored a cable's length from the Retvizan—probably the Diana—was struck at the water line and set on fire. Eighty persons on board perished.

The Russians sank two old steamers belonging to the Chinese Railway Company laden with stones at the entrance to the channel, in line with the lighthouse, and thus reduced the navigable way of the channel to less than 300 feet wide. This was done during the bombardment, the desperate operation being carried out under fire. The sunken steamers lay in the shape of a letter V.

On March 21 and 22 Admiral Togo made his fifth attack on Port Arthur, which was ineffectual, according to his own report, which was as follows:

"The combined fleet acted according to the plan arranged.

"Two flotillas of destroyers were outside Port Arthur, as instructed, from the night of the 21st until the morning of the 22d.

Although during this time our destroyers were under the fire of the enemy, they sustained no damage. The main fleet arrived off Port Arthur at 8 o'clock on the morning of the 22d.

"I dispatched a part of the fleet in the direction of Pigeon Bay, and ordered the battle-ships Fuji and Yashima to make an indirect bombardment against the inner side of the port. During the bombardment the enemy's ships gradually came out of the harbor, and at the time when the indirect bombardment stopped, which was about 2 o'clock, the number of Russian ships was five battle-ships, four cruisers, and several destroyers. We believed the enemy was trying, by making a movement of the fleet, to draw us near the forts. The enemy's ships shelled us indirectly, and many of the shots fell near the battle-ship Fuji, but our ships sustained no damage. About 3 o'clock our vessels withdrew off the port. Togo."

The next important move by Admiral Togo was on April 13, when Admiral Makaroff made a sortie and was defeated, with the loss of his own life and the destruction of his flagship, the details of which are told in another chapter.

As a result of this disaster to the Russian squadron all danger of attack upon Japanese transports bound for New Chwang was averted, and troops were poured into that place, which was captured with only slight resistance.

The first great disaster to Admiral Togo's squadron at Port Arthur occurred on May 15 and included the loss of the cruiser Yoshino and the battleship Hatsuse. The Yoshino collided with the cruiser Kasuga in a heavy fog and the latter sank. Only ninety of the crew was rescued, the remaining 210 going down with the ship.

The Hatsuse was sunk by a Russian mine. The battleship was cruising off Port Arthur, covering the landing of troops, when she struck a mechanical mine. Instantly she signaled for help, but the next moment struck a second mine that sealed her fate. Three hundred of the crew were saved by torpedo boats, but more than 400 were drowned, among them many of the minor officers.

On May 22 Vice-Admiral Skrydloff arrived at Port Arthur and assumed command of the naval operations left vacant by the death of Makaroff. The almost continuous assaults of Admiral Togo upon Port Arthur and Dalny compelled the Russians on May 26 to abandon the latter place. The Russian garrison at Dalny set fire to the place and fled to Port Arthur. They also destroyed a Russian gunboat. The Japanese occupied the town May 30 and found that the Russians had evacuated the place in such a hurry that they had failed to destroy much property. Over 100 barracks, the storehouses, railway and telegraph stations and 200 passenger and freight cars were uninjured. The big pier, however, was destroyed and the entrances to the docks were blocked with sunken steamers.

The Russian squadron at Port Arthur attempted to escape from the harbor on June 23, led by a steamer used for clearing mines. They were discovered by Japanese torpedo boats guarding the mouth of the harbor, and an engagement among the torpedo boats of both squadrons followed. Subsequently the Japanese decoyed the Russians out to sea, but before a general engagement could be begun the Russian ships made for the harbor. The Japanese torpedo fleet chased them, and in the attack damaged the battleships Peresviet and Sebastopol and the cruiser Diana. Eight separate attacks were made during the night, but the Russian vessels succeeded in returning to the harbor.

By June 30 the Japanese army in the rear of Port Arthur was within eight miles of the outer forts, when an assault was begun which resulted in the occupation of one mountain by the besiegers. and gave them a distinct point of vantage.

The decisive fight for the outer works was begun on July 26 and lasted for four days. The Russians occupied a line of trenches sixteen miles long. When the fog cleared on the morning of the 26th an attack began along the entire line and was kept up until dark. On the morning of the 27th it was resumed more fiercely than before and was concentrated on the right wing, commanded by Major General Kondratenko. The naval battery was made the target for the heaviest fire, as it was the most dangerous Russian position. At 9 o'clock the Japanese artillery fire slackened and the Japanese infantry advanced to the assault. For an hour the little brown men advanced intrepidly in the face of a fire which can only be described as 1,000 volleys in constant eruption. On July 30, after almost constant fighting, the Russians were forced to abandon their trenches and the Japanese occupied the first of the outer forts.

When the Japanese drove the Russians from their strong position on the last range of hills in front of the fortress by a surprise attack July 30, the Russians retired to the forts, but they also strongly held previously prepared advance lines to prevent the Japanese closing in on the fortified ridges. This line of forts was fourteen miles long, forming a semicircle from the east coast to four miles from the west coast and circling five miles northwest of Port Arthur. The fortress belt proper was a twelve-mile semicircle from coast to coast.

The Japanese position was a mile from the advance Russian line in the center of a range of hills called Fenghoano mountain. On their left flank was broken, hilly country east of Taku mountain.

Japs Checked in First Assault.

The keys to this Russian position were Taku and Shahku mountains, and these were taken early in August. Then the Japanese got siege guns and prepared for the first general assault August 20. The operations began in earnest at daybreak, August 20, with the bombardment of the whole line of Russian forts. Battery after battery of Japanese artillery was unmasked, the Russians apparently being unable to locate them, for they replied spasmodically. Under cover of the bombardment the Japanese infantry made a determined general advance against the Russians' first trenches, along the railroad, into the center of the Shuishi valley and also on the lines in the valley between the east and west of Shuishi village.

Six hundred yards south of the village the way into the city was protected by four advance half-moon forts in the form of a hollow square open at the rear and connected by bombproof trenches and having a twenty-foot moat in front. A desperate attempt was made by the Japanese in the afternoon to capture the strongest half-moon fort.

The infantry charged, cut the entanglements, crossed the moat with scaling ladders, stormed the fort and drove the Russians back. But the real strength of the enemy's position lay in the bombproof trenches extending south of the other half-moon forts. They were filled with troops and concealed machine guns, which poured a deadly hail of bullets into the Japanese and forced them to retire.

Russians Recapture Some Trenches.

About the same time the Russians recaptured the trenches in front of the redoubt in advance of the Rihlung fort.

On the right flank the Russian line was forced back. One Japanese regiment worked up the east slope and another regiment made its way up the west slope of One Hundred and Seventy-four Yards hill, under the concentrated fire of the Russian artillery, and captured the fort at the point of the bayonet at noon. The fort and hill were strongly supported by bomb-proof trenches, loop-holed and with sand-bag walls, and were armed with two four-inch guns, besides other guns. The Japanese captured five field guns and four machine guns. Their casualties were 1,400. Three hundred and fifty Russian dead were left in the position.

(The Siege of Port Arthur continued on page 421.)

CHAPTER XXV

SECOND ATTEMPT TO BOTTLE UP RUSSIAN FLEET

Admiral Togo Sends Four Merchant Ships Into the Channel Resulting in a Desperate Fight Between Torpedo Boats—Sinking of the Russian Torpedo Boat Silni—Lieutenant Krinizki, Her Commander, a Naval Hero—Makaroff Offers Battle With His Squadron Which Togo Declines—One of the Most Daring Naval Exploits of the War.

ADMIRAL TOGO'S second attempt to bottle up the Russian Port Arthur squadron resulted in a fierce fight between torpedo-boats and developed another naval hero in the person of Lieutenant Krinizki, commanding the Russian torpedo-boat Silni. The Japanese admiral sacrificed four merchant vessels, and, although he did not succeed in blocking the harbor channel, he inflicted considerable loss upon the enemy.

The daring attempt was made about 1 o'clock on the morning of March 27. The Japanese practically repeated the tactics of February 24 by sending in four fireships, preceded by a torpedo-boat flotilla, with the exception that the fireships this time were armed with Hotchkiss guns for the purpose of keeping off the Russian torpedo-boat destroyers.

The enemy's attempt was discovered by means of the shore searchlights and a heavy fire was opened from the batteries and from two gunboats which were guarding the entrance to the harbor. The Russian torpedo-boat destroyer Silni was outside on scouting duty, and to the dash and nerve of her commander, Lieutenant Krinizki, is chiefly due the defeat of the plans of the Japanese.

He at once made straight for the oncoming ships, under a hail of fire from the Hotchkiss guns, and torpedoed the leading ship, which sheered off, followed by the others, three of them being piled up on the shore under Golden Hill and one under the lighthouse.

The Silni then engaged the entire six torpedo-boats of the enemy, coming out from a terrific fight with seven killed and her commander and twelve of her complement wounded, but on the Japanese side only one boat's crew was saved.

The Japanese cruisers which supported the attack exchanged shots with the batteries and then drew off, after which Vice Admiral Makaroff took a steam launch and examined the fireships. An hour later the Japanese torpedo flotilla, followed by Vice Admiral Togo's fleet, came up from a southerly direction.

Just at daybreak Vice Admiral Makaroff, with his fleet, sailed out to engage the enemy, but after the ships' batteries had fired a few long-distance shots, Vice Admiral Togo decided to decline the issue and disappeared to the southward.

The news of the repulse of Vice Admiral Togo's second attempt to block Port Arthur created much rejoicing in the Russian capital, and among all classes the gallantry of the Silni and her commander was the subject of high praise.

Vice Admiral Makaroff sent the following telegram to the Emperor:

"I beg most humbly to report that at 2 o'clock this morning the enemy made a second attempt to block the entrance to the inner roadstead. For this purpose they dispatched four large merchant steamers, convoyed by six torpedo-boats, to the entrance.

"The enemy's ships were promptly discovered by the searchlights and were bombarded by the batteries and by the guardships Bohr and Otvajny.

"Fearing the enemy's ships might break through, Lieutenant Krinizki, commanding the guard torpedo-boat Silni, attacked the enemy and destroyed the bow of the foremost Japanese steamer with a torpedo. This steamer turned to the right and was followed by two others, with the result that the three were stranded to the right of the entrance. A fourth steamer went to the right of the enemy's ships and likewise sank to the side of the fairway. The Silni then battled with the enemy's six torpedo-boats. Engineer Artificer Swyereff and six seamen were killed and the commander and twelve seamen were wounded.

"At daybreak the enemy's battle-ship and cruiser squadrons appeared and I proceeded with the fleet under my charge to meet the enemy.

"The second attempt of the Japanese to block the entrance to Port Arthur has failed, thanks to the energetic defense by the sea and land forces, who acted as they did during the first attempt.

"The harbor remains perfectly clear.

"Makaroff."

In the following dispatch which Admiral Makaroff sent later, the "infernal machines" referred to were probably electric batteries intended to blow up the ships when they should reach a position in the channel which would interfere with the free passage of ships:

"I respectfully report that the enemy having withdrawn I returned to the harbor with the fleet.

"The torpedo-boat destroyer Silni, which stranded on a reef in consequence of damage caused to her engines by one of the enemy's shells, was floated during the course of the night and entered the harbor, thanks to the energy of her crew. Her commander, Lieutenant Krinizki, who was slightly wounded in the arm, did not quit his post.

"On the fireships were infernal machines, the wires connected with which were cut by Lieutenant Pilsoudsky of the irregulars, whom I dispatched on this task. They boarded one of the steamers as soon as it stopped, cut the electric wire and extinguished the fire, which would have lit up the entrance to the harbor to the enemy in the roadstead.

"In the morning a floating mine was found bearing an infernal machine, but the latter was successfully removed.

"The inspection made showed that the steamers utilized as fireships were not old. They were each of about 2,000 tons and they were armed with light caliber guns. Makaroff."

Viceroy Alexieff in a dispatch to the Emperor from Mukden said:

"During an attack by the enemy's fireships March 27 on the torpedo-boat Silni one of the latter's steam pipes and her steering gear were damaged, in consequence of which she was beached near Golden Hill, but since then she has been refloated. The number of men killed and wounded on the torpedo-boat has not yet been ascertained.

"Shortly before 5:25 this morning the enemy's torpedo-boats were sighted to the south of Port Arthur and the batteries opened fire on them. Toward 6 o'clock the enemy's squadron appeared on the horizon. Thirty batteries on Tiger Peninsula opened fire and our fleet steamed out of the harbor, the Bayan and the Askold leading, and also fired at the enemy. Our fire, however, ceased immediately in view of the great distance of the enemy's squadron.

"At 9:15 o'clock our entire squadron lined up in the roadstead. The Japanese squadron drew off in a southeasterly direction, evidently in order to avoid giving battle, and toward 10 o'clock it disappeared below the horizon. Alexieff."

Important additional details were contained in the following dispatch sent by General Smirnoff:

"Last night after moonrise the Japanese attempted to block the entrance to the harbor. Four ships were sent toward the port, convoyed by a torpedo flotilla. Toward 2:15 a. m. the approach of the enemy's ships was perceived by the guardships and batteries, which simultaneously opened upon them heavily.

"The fireships were preceded by torpedo-boats and followed at a considerable distance by larger ships, which opened on the forts, supporting the action of the fireships and torpedo-boats.

"Owing to the heaviness of our artillery fire and the boldness of our torpedo-boats, the fireships did not reach the entrance to the harbor. Two of them grounded on a reef under Golden Hill, another sank behind the first turn of the channel, struck by a torpedo from one of our boats, and the fourth sank, its bows touching a Japanese steamer sunk in the previous attempt off Majatschnaja Goroda. The entrance to the harbor remains clear.

"A Hotchkiss one-inch quick-firer was found aboard one of the sunken steamers, from which a fire had been kept up on our torpedo-boats.

"A boat left each of the sunken ships carrying their crews. One of these is believed to have been picked up.

"Toward 4 o'clock a. m. the enemy's torpedo-boats retired and the bombardment ceased. Vice Admiral Makaroff at once proceeded in a steam launch to inspect the enemy's sunken steamers.

"The enemy's torpedo-boats reappeared at 5 o'clock this morning. They were sighted south of Port Arthur and the batteries reopened on them. Toward 6 the enemy's squadron appeared on the horizon and ours steamed out to meet it. At 6:30 the batteries opened fire. Our ships' batteries soon ceased, the Japanese drawing off to the southeast, evidently declining an engagement. At 10 o'clock they disappeared below the horizon.

"Smirnoff."

Lieutenant Krinizki a Hero.

The credit for saving the Port Arthur squadron from being bottled up belongs to Lieutenant Krinizki, who displayed a heroism as great as any ever shown under equally adverse conditions. In support of this opinion the writer quotes Rear Admiral Ingles, of the British navy. In writing of this event Admiral Ingles says:

"Fortunately for the security of Port Arthur there was such a man in Lieutenant Krinizki, a young officer in command of one of the newest thirty-two knot destroyers built at St. Petersburg.

In his little ship of 350 tons he charged out on the enemy with a courage worthy of all praise.

"If the official story is the whole truth this young officer undertook the task of saving Port Arthur from being corked single handed and succeeded. It was an enterprise in which the odds were heavily in favor of entire annihilation, and at best it meant death for many of the crew.

"This lieutenant was on guard as the enemy's dummies and half a dozen torpedo craft drew near, and when he saw that the guns ashore were making no impression he leaped forth from his position of comparative safety. He first succeeded in so damaging the foremost merchant ship that the two following ones were sent off their course in confusion, while a fourth sank on the left hand side of the fairway, all of them being wrecked on the outside of the narrows of the harbor.

"This intrepid young officer achieved his purpose, but only to land himself in a hornet's nest. He is represented to have fought six Japanese torpedo-boats without any assistance, although there must have been at least nine sister vessels close at hand. The Silni has only one twelve-pounder and three three-pounders, and with these weapons its brave commanding officer engaged in a most gallant and hopeless but glorious struggle. He and twelve of his seamen were wounded and the chief engineer and six sailors were killed.

"If the facts are as set out in a short dispatch to St. Petersburg the name of Lieutenant Krinizki deserves to be placed on the roll of heroes. When the din of the present struggle has died away his exploit should remain a cherished memory of the Russian fleet. It is good news that so brave a man escaped with his life, though his frail craft was so damaged that after retiring it sank."

Large crowds gathered before the bulletin boards in St. Petersburg to read the official accounts of the second unsuccessful attempt of the Japanese to bottle up the Russian fleet. Lieutenant Krinizki's heroic attack on the enemy's fireships and torpedo-boats and the action of the three Russian officers in boarding the

COSSACK LEADER RECEIVING ORDER FROM COMMANDER.

EARLY tn the war every available Cossack was ordered to the front, as upon the wonderful Cossack cavalry was to fall the hardest tasks of the land campaign. The scene represents a temporary Cossack station, the tattered war flag leaning against the gate, and the striped posts as well as the two sturdy guards, indicating that here is the headquarters of the commandant and staff.

COOLIES AT WORK UPON THE PORT ARTHUR DEFENSES.

THE repeated bombardments directed by the Japanese against the defenses of Port Arthur served only to stimulate the Russians in their efforts to make them impregnable. In this work the Russian whip and the Chinese cooley played a leading part.

burning steamers, extinguishing the flames and cutting the wires connecting with the infernal machines, evoked enthusiasm, while the almost universal exclamation applied to Vice Admiral Makaroff, as the people turned away, was "molodetz," which might literally be translated, "he's a dandy."

The Emperor telegraphed to Vice Admiral Makaroff an order decorating the officers and men of the torpedo-boat destroyer Silni with the St. George's cross.

Admiral Togo's report of his sixth attack on Port Arthur and second attempt to bottle up the Russian fleet is modest, direct and businesslike, as are all of his communications. The report is as follows:

"About 3:30 a. m. of the 27th of March the 'bottling up squadron,' composed of four ships, escorted by a torpedo-boat destroyer flotilla and torpedo-boat flotilla, reached outside of Port Arthur and without minding the searchlights of the enemy steered straight towards the entrance of the harbor. About two marine leagues from the entrance the 'bottling up squadron' was discovered by the enemy. Thereupon the shore batteries and guardships showered hot fires upon the squadron, but, in spite of the terrific fire, the ships made their way into the inner roadstead, one after the other.

"The steamer Chiyo Maru anchored at a position about half a cable from the Golden Hill, blew up itself and sank. The Fukui Maru passed a little ahead of the Chiyo Maru by its left side and at the moment when it was lowering anchor was shot by a torpedo from the enemy's destroyers and sank in that position. Hachi-Hiko Maru anchored to the left of the Fukui Maru and blew up itself and sank.

"Yoneyama Maru, colliding with the stern of one of the enemy's torpedo-boat destroyers, passed between Chiyo Maru and Fukui Maru and anchored in the middle of the roadstead. At this moment the ship was shot by a torpedo from the enemy, and, consequently by reason of that torpedo, it was carried toward the left shore and sank sideways.

"The result of the action being as above described, there is some space left between Hachi-Hiko and Yoneyama Maru. It is a matter of regret that the roadstead could not be completely closed up. The casualties were as follows:

"Killed—Commander Hirose Takeo, one under officer, and two sailors.

"Seriously wounded—Sub-Lieutenant Masaki, Engineer Kurita and six sailors.

"The remainder were safely taken in by our torpedo-boat destroyer flotilla and torpedo-boat flotilla.

"Of the torpedo-boat flotilla, the Oadaka and the Tsubame, while escorting the 'bottling up squadron' and about one mile from the entrance of Port Arthur, engaged in a fight with one destroyer of the enemy and inflicted serious damage on it.

"Although both our destroyer flotilla and torpedo-boat flotilla were subjected to terrific firing from the enemy until dawn, not the slightest damage was done to any of the boats."

On May 3 Admiral Togo made another daring attempt to seal the channel at Port Arthur. When the vessels were all ready to make the attempt a wild storm suddenly broke and scattered them in every direction, but later the Admiral got them together and started for the entrance to the harbor. The blockading flotilla consisted of eight stone laden merchantmen. They were escorted by the gunboats Akagi, Commander Fugimoto; the Choikai, Commander Iwamura; the second torpedo boat destroyer flotilla under Commander Shida, the third flotilla under Commander Tsuchiya, the fourth under Commander Nagai, the fifth under Commander Mano, the ninth under Commander Yashima, the tenth under Commander Otakiand, the fourteenth under Commander Sakurai.

The attack exceeded all of its predecessors in the desperation and courage displayed by the volunteer fleet.

Commander Hayashi, who was in command, boldly ran his merchant ships into the mouth of the harbor in the face of a terrific fire. The Russians had fires burning at the harbor mouth and also used powerful searchlights. They poured an incessant fire into the channel. Lieutenant Sosa, who was in command of one of the stone laden ships, forced his vessel into the channel, rammed his way through the booms and reached the center of the inner entrance. Here he anchored his vessel and then blew it up. It sank immediately. Six other stone ships advanced to the mouth of the harbor. The Russian fire now became highly effective, and a number of sunken mines, struck by the vessels, began to explode. These explosions caused heavy casualties among the Japanese. Six of the merchantmen were sunk inside the mouth of the harbor, some by Russian guns and the others by mines. It was believed at the time that the channel had been completely blocked, but this proved not to be the case, although the passage of warships was seriously interfered with for a long time.

In his report of this affair Vice Admiral Togo says: "This undertaking, when compared with the last two, resulted in greater casualties to our side. Owing to the inclemency of the weather and the increased preparation for defense completed by the enemy, we could save none of the officers or crews of four of the merchant ships. It is regretted that nothing could be learned of their brave discharge of their duties, but the memory of their exemplary conduct will remain long in the imperial navy. The destroyer and torpedo boat flotillas, besides resisting the enemy, bravely fought against wind and waves. The torpedo boat flotilla approached close to the mouth of the harbor and rescued more than half the men.

"The third detachment, Rear Admiral Sewa commanding, reached Port Arthur at 6 o'clock the same morning. The first detachment, under Vice Admiral Togo and Rear Admiral Nashiha, arrived off Port Arthur at 9 in the morning for the purpose of protecting the destroyers and torpedo boat flotillas and to search for the crews of the sunken merchantmen. The vessels remained until late in the evening, but their search was fruitless. The day was foggy and it was impossible to observe the condition of the enemy."

Two torpedo boats and one destroyer were badly damaged in this attempt to block the harbor.

The officers and men in this daring enterprise were all volunteers, and those that lost their lives went to their deaths as cheerfully as if they had been embarking upon a pleasure excursion. Even the Russian reports of the affair contained glowing tributes to the daring and reckless bravery of the Japanese invaders, which probably is not excelled in the naval history of any nation.

CHAPTER XXVI

GALLANT DEATH OF ADMIRAL MAKAROFF

Russian Naval Commander at Port Arthur Goes to His Death on the Battleship Petropavlovsk, Which is Blown up by a Japanese Mine—Crew of 525 Men go Down and but 80 are Rescued—Verestchagin, the Great Russian War Painter, Also Perishes—Makaroff, Called "The Cossack of the Sea," One of the Most Dashing and Ablest Officers in Russian Navy—Mourned by Russia and Japan.

THE most dramatic and the crowning tragedy of the long siege of Port Arthur was the gallant death of Vice Admiral Makaroff, who went down with his flagship, the Petropavlosvk, on the night of April 13. In the midst of a fierce battle at sea the Petropavlosvk, one of the first class battleships of the Russian navy, struck a mine planted in her sea path by Admiral Togo's squadron and sank with her crew of 525 men, of whom only eighty were rescued.

Besides Vice Admiral Makaroff, whose body was almost blown to pieces and for a time lay weltering in its own blood on the deck of his flagship, Vasili Verestchagin, the famous Russian painter of war scenes, was also killed, and the Grand Duke Cyril seriously wounded.

Tuesday night, April 12, Vice Admiral Makaroff took to sea with his entire squadron, including fourteen torpedo-boats. The next night, April 13, in the teeth of a gale, eight torpedo-boats were sent out to reconnoiter. From Golden Hill through the blackness the searchlights of the fortifications could be seen flashing over the inky waters of the roadstead and far out to the hazy horizon.

At 11 o'clock observers on Golden Hill heard firing at sea and counted seven shots, but could see nothing. At daybreak through the light haze to the southward, about five miles from shore, six torpedo-boats were seen strung out in line, and all firing. In the lead and outstripping the others was a boat heading at full speed directly for the entrance of the harbor. The last in the line was beclouded in steam and lagging. She had evidently been hit. It was difficult to distinguish the Russian ships, but finally it was seen that the leader and the laggard were both Russian and that the four others were Japanese.

The flash of the guns and the splash of the projectiles as they struck the water showed the intensity of the conflict.

Torpedo Destroyers in Terrific Fight.

The torpedo-boat from which steam was escaping was firing viciously. The four center craft drew together, concentrating their fire upon her, but the crippled destroyer poured out her fire and was successfully keeping off her assailants.

The signal station flashed the news to the men of the batteries that the vessel was the Strashni.

The unequal combat was observed with breathless interest, but the net drew close around the doomed boat. The four Japanese vessels formed a semicircle and poured in a deadly fire. The steam from the Strashni grew denser, covering her like a white pall. Still she fought like a desperately wounded animal brought to bay.

Running straight for the adversary barring her way to safety she passed the Japanese astern and fired at them.

At this stage Vice Admiral Makaroff, who had been observing the progress of the conflict through a telescope, signalled to the cruiser Bayan, lying in the inner harbor, to weigh anchor and go out to the rescue.

The Japanese destroyers clung to their victim like hounds in a chase. They had become separated, but again resumed their formation.

Small jets of flame and smoke were spurting from the light rapid-firers, varied by denser clouds as torpedoes were discharged against the Strashni.

Sinking of the Strashni.

It was the end. The stricken boat loosed a final round, but it was as if a volley had been fired over her own grave, for she disappeared beneath the waves, only a little cloud of steam marking the place where she went down.

Satisfied with what they had accomplished, the Japanese torpedo-boats turned and made off at full speed, followed by the Bayan. To their support came six of the enemy's cruisers. Still the Bayan went on, seemingly inviting certain destruction. She soon ported her helm to bring a broadside to bear upon the foe, advancing in line of battle. She opened on some of them and turned quickly and stood on into the hail of the enemy's broadsides. The Japanese steamed at a slight angle, enabling all their guns to bear and projectiles rained around the Bayan, raising columns of water as they burst, but none struck home.

To the eastward suddenly appeared five more Russian torpedo-boats returning to the harbor under forced draught. Two of the Japanese cruisers were immediately detached to cut them off, but the Bayan, noticing the movement, checkmated it by turning a hot fire upon them. The movement was effective. The Japanese cruisers slowed down and the torpedo-boats slipped through into the harbor.

Admiral Makaroff Goes Out to Battle.

Meantime, in accordance with Vice Admiral Makaroff's order, the battle-ships and cruisers in the inner harbor slipped anchor. Majestically the Petropavlovsk, flying the admiral's flag, steamed through the entrance. On her appearance the more formidable enemy of Japanese cruisers turned and fled. The admiral signaled the Bayan to return. Then a stream of varicolored signal flags fluttered out "Bravo Bayan."

By this time the entire Russian squadron was in the outer harbor. Besides the Petropavlovsk there were the battle-ships

Peresviet, Poltava, Pobieda and Sevastopol, the cruisers Novik, Diana and Askold and the torpedo-boats. The flags announcing the admiral's approbation of the Bayan were hauled down and replaced by another signal. Immediately the torpedo-boats dashed ahead and the heavier ships began to spread out.

Big Guns of Battle-ship Roar.

Seeing the flight of the Japanese cruisers, the Petropavlovsk opened fire with her great guns, but the enemy were out of range and soon disappeared.

The Russian squadron continued the chase, finally fading from view.

In about an hour it came back in sight. Far beyond it, the number of points from which smoke arose announced the presence of the Japanese. Nearer and nearer came the vessels toward Port Arthur, and at last behind the Russian squadron came a fleet of fourteen, of which six were battle-ships and the remainder armored and unarmored cruisers flying the flag of the Rising Sun.

Unable to get within effective range of Vice Admiral Makaroff's ships, the Japanese stopped about eighteen versts (twelve miles) from shore.

Russians Reach the Harbor.

The Russian squadron, with the Patropavlovsk leading, arrived at the entrance to the harbor and drew up in line of battle. Another signal was floated from the flagship, and the torpedo-boats at once proceeded through the entrance into the inner harbor. Vice Admiral Makaroff was evidently unwilling to risk his vulnerable craft to the heavy projectiles of the enemy's armored ships.

The observers on Golden Hill watched the Petropavlovsk as she steamed toward Electric Cliff. The frowning marine monster, whose guns were ever turning toward the enemy, was prepared to send huge messengers of death against him.

All was quiet. It was the hush before a battle—the hush when every nerve is strained to get into impending danger.

The Japanese ships were without movement, save that caused by the heaving sea.

Petropavlovsk Blown Up.

The Petropavlovsk was almost without headway when suddenly she trembled. She seemed to rise out of the water, a tremendous explosion rent the air, then a second and then a third. Fragments flew in all directions and wreckage and men were mixed up in a terrible mass.

One could hardly realize the horror of it when the ship began to list. In a moment the sea seemed to open and the water rushed over her. The Petropavlovsk had disappeared.

The floating woodwork and the few men struggling in the water were all that was left to recall the splendid fighting machine which a few hours before had sailed out of the harbor.

The same shock experienced by the observers on Golden Hill paralyzed for a moment the men on the ships, but when it passed torpedo-boats and small boats hastened to the rescue of the survivors.

Eager to ascertain what had occurred on board the sunken ship, the watchers on Golden Hill rushed to a landing where a small remnant of the gallant crew were being put ashore and conveyed to a hospital. Signalman Pochkoff, who was slightly wounded, was able to give a remarkably clear statement of the disaster.

"We were returning to the harbor, the Petropavlovsk leading," he said. "Some of our cruisers which had remained in the harbor came out and steamed toward the enemy, firing sixteen shots at him with their bow guns. They then retired. The enemy numbered fourteen heavy ships, nearly all armored, while ours were nine. Against their armored cruisers we had only the Bayan. I stood in the wheelhouse on the bridge of the Petropavlovsk, looking up the signal book. The admiral's last signal had been for the torpedo-boats to enter the harbor.

"The Petropavlovsk slowed speed and almost stood still. Suddenly the ship shook violently. I heard a fearful explosion, immediately followed by another and then another. They seemed to me to be directly under the bridge. I rushed to the door of the wheel-house, where I met an officer, probably a helmsman. I could not pass him and I sprang to the window and jumped out. The ship was listing and I feared that every moment she would turn over.

Makaroff Dead on the Deck.

"On the bridge I saw an officer weltering in blood—it was our admiral—Makaroff. He lay face downward. I sprang to him, grasped him by the shoulder and attempted to raise him.

"The ship seemed to be falling somewhere. From all sides flew fragments. I heard the deafening screech and the frightful din. The smoke rose in dense clouds and the flames seemed to leap toward the bridge where I was standing beside the admiral. I jumped on the rail and was washed over, but succeeded in grabbing something.

"On our ship was an old man with a beautiful white beard, who had been good to our men. He had a book in his hand and seemed to be writing, perhaps sketching. He was Verestchagin, the painter."

The Petropavlovsk began to settle slowly by the head, heeling far over to starboard, until her rail was under water, her bow disappeared and then the foremast sank, but the conning tower could still be seen. Then her smokestacks disappeared. They seemed to fall through the water and the sea began to engulf the mainmast. Her after turret, with its guns pointing skyward, quickly vanished; then her stern, with the port propeller still revolving. Figures could be seen desperately clutching at the slippery hull and tongues of flame. A last explosion and all was over. The flagship was no more.

Boats from the cruiser Gaydamak hurried to the scene of the disaster. It was fifty minutes past 9 in the morning.

The Pobieda also struck a Japanese mine and sustained damages, but was able to regain the harbor.

The Russian Report.

The text of Viceroy Alexieff's report to the Czar concerning the Petropavlovsk disaster and the torpedo-boat engagement which preceded it is as follows:

"I respectfully report to your Majesty that April 11 the whole effective squadron at Port Arthur sailed out six miles to the southward to maneuver and toward evening returned to port. April 12 a flotilla of eight torpedo-boat destroyers were out to inspect the islands, having received orders to attack the enemy should he be encountered in the course of the night.

"Owing to the darkness and a heavy rain three of the destroyers became separated from the flotilla and two of them returned to Port Arthur at dawn. The third, the Strashni, having, according to the evidence of her seamen, encountered several Japanese destroyers, took them in the darkness for Russian ships, and, giving the signal of recognition, joined them at dawn. She was recognized by the enemy and there was a fight at close quarters, in which her commander, midshipman and engineer and most of her crew were killed. Maleiff, her lieutenant, although wounded, continued firing on the enemy.

"At dawn April 13 the cruiser Bayan went out, preceded by destroyers, and hurried to the rescue. About sixteen miles from Port Arthur the Bayan saw the destroyer Strashni engaged with four Japanese destroyers. Shortly afterward an explosion occurred and the Strashni sank. Driving off the enemy's destroyers with her fire, the Bayan approached the scene of the fight, lowered her boats and had time to save the remnant of the destroyer's crew. Unfortunately only five men were swimming. Their lives were saved.

"The cruiser was obliged to fight on her starboard side with six Japanese cruisers which came up. Having picked up her boats the Bayan regained the harbor, suffering no damage or loss, although covered with fragments of shells.

"The cruiser Diana and five destroyers hastened to her succor and at the same time the other cruisers, the battle-ships Petropavlovsk and Poltava and some destroyers came out from the roadstead and the other battle-ships left the harbor in column formation, with the Bayan at the head and the destroyers on the flank. Vice Admiral Makaroff proceeded to the scene of the fight, whither more Japanese destroyers and cruisers were going.

"After a short fusillade at fifty cable lengths (1,000 yards) the ships drew off.

"A squadron of nine Japanese battle-ships appeared at 8:40 a. m. and our ships retired toward Port Arthur. In the roadstead they were joined by the battle-ships Pobieda, Peresviet and Sevastopol, which were coming out through the channel. The squadron was drawn up in the following order: Askold, Bayan, Diana, Petropavlovsk, Peresviet, Pobieda, Novik, five destroyers and two torpedo cruisers. They turned toward the left, but when approaching the mouth of the channel the destroyers were signaled to return to the harbor and the cruisers to proceed. Maneuvering, with the Petropavlovsk at its head, the squadron turned to the east, making toward the enemy on the right.

"At 9:43 a. m. an explosion occurred at the right side of the Petropavlovsk; then a second and more violent explosion under her bridge. A thick column of greenish-yellow smoke was seen to rise from the battle-ship, her mast, funnel, bridge and turret were thrown up and the battle-ship heeled over on her starboard side. The Petropavlovsk was surrounded by flames and in two minutes sank bow first.

"Some of her crew escaped. The cruiser Gaydamak, which was a cable length away, lowered boats and succeeded in rescuing Grand Duke Cyril and forty-seven seamen. The destroyers and boats from the Poltava and Askold also picked up some of the

Petropavlovsk's crew. Altogether seven officers and seventy-three men were saved. The Poltava, which was following the Petropavlovsk two cable lengths astern, stopped her engines and remained on the scene of the disaster.

"At a signal from Rear Admiral Ouktomsky, the other war ships made for the entrance of the harbor, maneuvering toward the Peresviet in line. A mine exploded under the starboard side of the Pobieda. She listed, but proceeded and entered the harbor with all the other ships astern of her. The enemy remained in sight until 3 o'clock, and then disappeared.

"The night preceding the sortie of the squadron lights and outlines of ships were seen in the distance from the roadstead, and the commander of the fleet kept watch in person until dawn from the cruiser Diana, stationed in the outer roadstead. He left her at 4 o'clock in the morning.

"In concluding, I take the liberty to announce respectfully that despite the ill success which has attended the Pacific fleet the crews of the ships retain their morale and are ready to perform all duties required of them. The gracious words of your Majesty addressed to the seamen at this painful hour of trial serve as a consolation and a support to all the force in their efforts to overcome the enemy, to the glory of their beloved sovereign and their country. Alexieff."

The official report of Admiral Togo throws additional light on the details of the fight and enables the reader to view from the Japanese side. It is as follows:

"The fourth and the fifth destroyer flotillas, and the fourteenth torpedo flotilla, and the Keryo Maru reached the mouth of Port Arthur at midnight of the 12th and effected the laying of mines at several points outside of the port, defying the enemy's searchlight.

"The second destroyer flotilla discovered at dawn of the 13th one Russian destroyer trying to enter the harbor, and after ten minutes' attack sunk it. Another Russian destroyer was discovered coming from the direction of Laotishan and was attacked, but it managed to flee into the harbor. On our side no casualties except two seamen on the Ikatsuchi slightly wounded.

"There was no time to rescue the enemy's drowned crews, as the Bayan approached. The third fleet reached outside of Port Arthur at 8 a. m., when the Bayan came out and opened fire. Immediately afterward the Novik, Askold, Diana, Petropavlovsk, Pobieda and Poltava came out and made an offensive attack.

"Our third fleet, hardly answering and gradually retiring, enticed the enemy fifteen miles southeast of the port, when our first fleet, informed through wireless telegraphy from the third fleet, suddenly appeared before the enemy and attacked them.

"While the enemy was trying to regain the port a battleship of the Petropavlovsk type struck mines laid by us the previous evening and sunk. Although another ship was observed to have lost freedom of movement, the confusion of the enemy's ships prevented us from identifying it. They finally managed to regain the port.

"Our third fleet suffered no damage, and the enemy's damage beside above mentioned probably slight also. Our first fleet did not reach firing distance.

"Our fleets retired at 1 p. m., prepared for another attack. They resailed April 14 toward Port Arthur. The second, the fourth and the fifth destroyer flotillas and the ninth torpedo flotilla also joined at 3 a. m. and 7 a. m. No enemy's ship was found outside of the port. Our first fleet arrived at the port at 9 a. m., and discovered three mines laid by the enemy, and destroyed all.

"The Kasagi and Nisshin were dispatched to the west of Laotishan, and made an indirect bombardment for two hours, it being their first action. The new forts on Laotishan were finally silenced. Our forces retired at 1:30 p. m."

In concluding his report, Admiral Togo says:

"The fact that not a single man was seriously injured in these successive attacks must be attributed to his Majesty's glorious virtue. The officers and men proved gallant and did their utmost in the discharge of their duties, despite many things that seemed beyond human control.

"The ships freely moving over the enemy's ground without suffering any damage must be attributed to heavenly assistance."

Japanese Strategy and Daring.

The success of the system of placing deadly counter-mines by the Japanese was due largely to a series of careful observations made by the Japanese during their previous attacks on Port Arthur. The Japanese saw the Russian fleet leave the harbor and return to it several times, and they discovered that the Russian warships followed an identical course every time they came out or went in, evidently for the purpose of avoiding their own mines.

The Japanese took bearings on this course. When the destroyer divisions of the Japanese torpedo flotilla laid the counter-mines during the night of April 12-13 they placed them along this course. The laying of these counter-mines was exceedingly perilous, because if any Japanese boat with mines on board had been struck by a lucky Russian shot she would have been annihilated.

The weather of the night of April 12-13 favored the work. There was a heavy rain, the night was dark and cloudy and the Russian searchlights playing over the channel failed to reveal the presence of the Japanese destroyers.

Rear Admiral Dewa was in command of the Japanese squadron which decoyed the Russian ships over the field of mines. His squadron consisted of the cruisers Chitose, Yoshino, Kasagi and Takasago, all unarmored vessels, which presented a tempting bait for the heavier Russian ships.

Vice Admiral Togo directed the flank attack. He had the battle-ships Hatsuse, Mikasa, Asahi, Shikishima, Yashima and Fuji. He waited thirty miles out at sea until Rear Admiral Dewa signaled him by wireless telegraphy to come in. His vessels then dashed at full speed toward the entrance of the harbor. All the battle-ships under Vice Admiral Togo are capable of a speed of eighteen knots and they quickly covered the distance.

It is not clear what warned the Russians that they had been trapped, but they probably discerned the battle-ship squadron on the horizon and retreated precipitately to the harbor. Vice Admiral Togo did not succeed in preventing the Russians from entering, but did force them to a disastrous retreat, which ended in the destruction of the Petropavlovsk and the disabling of the Pobieda.

Makaroff, the "Cossack of the Sea."

Vice Admiral Makaroff, the active commander of the Russian squadron, who lay dead on the deck as his flagship sank, was one of the most daring and popular officers in the Russian navy. He had already shown his high courage and fighting tactics in the torpedo-boat fight which he commanded from the cruiser Novik.

His ideas ran counter to those generally accepted by naval experts and strategists. One of his pet aversions was armor. The admiral was ever outspoken in his criticism of the heavy battle-ship, which he always had contended placed too many eggs in a single basket, and showed his partiality for a fleet composed of a large number of swift unarmored cruisers armed with heavy guns and attended by an unlimited number of torpedo boats. The protected cruiser Novik, to which he transferred his flag in the attempt to rescue a torpedo-boat destroyer, is practically without armor protection.

Admiral Makaroff's predilection for swift ships had earned him the sobriquet of the "Cossack of the Sea."

He began his career with several gallant exploits during the Turko-Russian war, while commanding a steamer which he had rigged up as a torpedo-boat, and played havoc with the sultan's flotilla in the Black Sea and Danube.

Special honors were done his memory both in St. Petersburg and Tokio. In the former city a requiem mass was said for him, attended by the Czar and Czarina. The Admiralty Church was

MODERN OFFICERS OF JAPAN.

WITHIN the past twenty years the Japanese Army has adopted features from all the great nations. The general officer's uniform is distinctly French; the cavalry officer wears the flat German cap.

MODERN INFANTRY OF JAPAN.

THE soldiers appear in white summer uniform, with French trousers and gaiters, and German field cap, knapsack, coat, and cartridge pouch. The sword-bayonet is both French and German.

WANDERING MUSICIANS PLAYING TO JAPANESE BEAUTY.

In all the so-called holy cities of Japan are found wandering musicians, bound by vows to follow this life of minstrelsy. Their chief instruments are the flute and samisen, and they helped to pass away the time of many a Japanese beauty whose husband or lover was at the front.

filled and 20,000 people stood outside with bared heads, and wept while the service was in progress.

Japanese Honor Makaroff's Memory.

In Japan funeral processions were organized in many cities and a day of abstinence was observed in the Japanese navy, just as if a Japanese hero had passed away.

Speaking for the naval staff, Commander Ogasawara published a lengthy statement in which he lamented the death of the Russian vice admiral and pronounced it to be a loss to the navies of the world. Commander Ogasawara reviewed the life, professional career and the personal attributes of Vice Admiral Makaroff and declared that he was entitled to be classed with the best admirals in the world.

Verestchagin and the Spanish War.

Vasili Verestchagin, the famous Russian artist, who perished with Vice Admiral Makaroff on the flagship Petropavlovsk, was the guest of the admiral and was engaged in securing subject matter for a cycle of war studies.

He was pre-eminent in the world of art. His incomparable genius and passionate realism gained him equal fame throughout Europe and America. Verestchagin came from a noble house of Russia. He was born at Tcherepovets, in Novgorod, in 1842, and graduated from the Russian Naval Academy when 17 years of age. He was a pupil of the famous Gerome at Paris for many years, and under the tuition of the latter developed the dominating ambition of his life, which was to become the delineator of the horrors of war. Verestchagin was a lover of peace, and hated warfare, which he characterized as the "reversal of Christianity."

He purposed to impress upon the peoples of earth the barbarity of war through the medium of his realistic canvases, and thus to admonish them to peace. He served through the Russian campaign with Kauffman in 1867, and through the Russo-Turkish conflict in 1878, and subsequently exhibited a cycle of twenty paintings depicting the principal engagements. His paintings of the Spanish-American struggle are familiar to the public in this country, the best remembered of them being "The Charge of Roosevelt's Rough Riders Up San Juan Hill" and "Lawton's Last Fight."

Sunk By a Submarine.

The loss of the Russian battleship Petropavlovsk, with Admiral Makaroff, nearly all his staff and her crew of 700 men, will be memorable in history, not only because of the loss of life, but also, it is now believed, because she was the first vessel to be destroyed by a submarine boat.

Viewing the Port Arthur disaster as the result of a submarine attack everything is clear—the peculiar maneuvering of Admiral Togo's fleet, the seeming stupidity of the Russians, the injury to the Pobieda, the startling rapidity with which the Petropavlovsk sank and the belief of the Russians that their battleship had struck a mine.

In the same engagement in which the Petropavlovsk was sunk the first-class battleship Pobieda was struck by some agency unknown to the Russians and disabled, her stern being almost blown off. Her commander is one of those who holds to the theory of the Japanese submarine. He says he was in clear water at the time, and there were two explosions separated by only an instant of time. These almost keeled his vessel over. There were no contact mines near, and the fixed mines could not be set off except from the shore. There was not a Japanese torpedo boat within two miles of her on the surface.

There were two shocks in the case of the Petropavlovsk as in that of the Pobieda. This is taken by the naval experts mentioned to indicate concerted attack by two submarines or extremely rapid firing by one.

CHAPTER XXVII

THE JAPANESE MARCH TO THE YALU

In Eight Weeks the Mikado's Land Forces Sweep the Russians Out of Korea and Across the Yalu, Practically Holding all the Territory in Dispute at the Beginning of the War—One of the Most Marvellous Marches in History—Japanese Defeat Cossacks at Chong-Ju—General Kuropatkin in Command of the Russian Forces.

THE complete occupation of Korea by the Japanese army and the advance to the Yalu river, on the Manchurian side of which stream the first great battle of the war was fought, constitutes one of the greatest and most astonishing military movements in the history of the world.

Within eight weeks from the outbreak of the war the Japanese army landed at Chemulpo, Korea, marched west thirty-five miles to the capital, Seoul, established a military base there, swung to the left, marched 150 miles to Pingyang, fortified it and created a base, extended its lines from Pingyang to Gensan, pressed north to Chongju, where it encountered a Russian Cossack force and drove it out, and continued its march to Wiju, at the mouth of the Yalu River.

In the 300 mile march from Seoul to Wiju there were numerous small skirmishes with Russian scouts and outposts, but the Russians were not in force anywhere in Korea and hence there was no great battle on the peninsula. The most serious clash was at Chongju, where the Japanese completely routed the Russians.

Although there was little fighting, the march across the Korean peninsula was a bitter hard one. At the beginning the roads were frozen and deep in snow, and later on they were knee deep in mud. Over such roads the Japanese army averaged six miles a day and carried its own provisions and dragged its artillery.

Before any serious skirmishes were fought the command of

the Russian land forces was intrusted to General Kuropatkin, former minister of war, and a man of established military reputation. His first official report was of a considerable engagement at Chong Ju on March 28, in which the Russians lost heavily and were driven toward the Yalu. This engagement technically may be called the first battle of the war, as the number of troops engaged would justify that designation. Cavalry and infantry on both sides were engaged. The Japanese fought gallantly, but were so exhausted that they were unable to follow up the retreating Russians, who fell back in good order on Kasan.

The battle came as the climax of three days of skirmishing between the outposts of the hostile armies confronting each other between the Yalu and Ping-Yang.

The engagement was on the sixth anniversary of the Russian occupation of Port Arthur. The operations took the form of a cavalry attack by six companies of Cossacks, led personally by General Mishtchenko, against four squadrons of Japanese cavalry, which the general believed to be beyond Chong-Ju, but which he found to be in occupation of that town.

Despite a cross-fire which General Mishtchenko cleverly directed against the enemy he pays a tribute to their tenacity and bravery, the Japanese only ceasing to fire after a combat which lasted for half and hour. Before the Russians could follow up their advantage three Japanese squadrons galloped toward the town, which two of them succeeded in entering, while the third was driven back in disorder, men and horses falling. The fire maintained on the town was so destructive that the Japanese were unable to make an effective return. Further Japanese re-enforcements arrived an hour later and General Mishtchenko was forced to retire.

Kuropatkin Tells of the Battle.

General Kuropatkin's report to the Czar was as follows:

"I have the honor respectfully to communicate to your Majesty the report of General Mishtchenko, dated March 28, at 10

p. m., which says: 'For three consecutive days our small outposts attempted to draw the Japanese cavalry into action, but their patrols, after contact was established, retired beyond Chong-Ju, about fifty miles northwest of Ping-Yang.

" 'Having learned that four squadrons of the enemy were posted five versts beyond Chong-Ju, on March 27, six companies marched toward Kasan and on March 28 reached Chong-Ju at 10:30 a. m. As soon as our scouts approached the town the enemy opened fire from behind the wall. Two squadrons promptly dismounted and occupied the heights, 600 yards distant. An engagement ensued.

" 'In the town a company of infantry and a squadron of cavalry were lying in ambush. Our men were re-enforced by three companies and attacked the Japanese with a cross fire. Notwithstanding this and our commanding position the Japanese gallantly held their ground, and it was only after a fierce fight of half an hour's duration that the Japanese ceased fire and sought refuge in the houses. The Japanese hoisted the Red Cross flag at two points.

" 'Soon afterward three squadrons of the enemy were seen advancing along the Kasan road at full gallop toward the town, which two of the squadrons succeeded in entering, while the third fell back in disorder under repeated volleys from our troops. A number of men and horses were seen to fall.

" 'For an hour afterward our companies continued to fire on the Japanese in the town, preventing them from leaving the streets and houses.

" 'An hour and a half after the beginning of the engagement four companies were seen on the Kasan road hastening to attack. I gave the order to mount, and the entire force, with a covering squadron, advanced in perfect order and formed in line behind the hill. The wounded were placed in front and the retirement was carried out with the deliberation of a parade.

" 'The Japanese squadron which was thrown into disorder was evidently unable to occupy the hill which we had just evacuated, and their infantry arrived too late.

" 'The detachment protecting our rear guard arrived quietly at Kasan, where we halted for two hours in order to give attention to our wounded. At 9 p. m. our force reached Noo-San.

" 'It is supposed that the Japanese had heavy losses in men and horses. On our side, unfortunately, three officers were severely wounded—Stepanoff and Androoko in the chest and Vaselevitch in the stomach. Schilnikon was less seriously wounded in the arm, but did not leave the field. Three Cossacks were killed and twelve were wounded, including five seriously.'

"General Mishtchenko bears witness to the excellent conduct and gallantry of the officers and Cossacks, and especially praised the Third company of the Argunsk regiment, commanded by Krasnostanoff.

Kuropatkin."

Chong-Ju, because of its superior natural surroundings, was the strongest place between Ping-Yang and Wiju. Besides these natural advantages there was an old Korean fort there, which, had it been defended with spirit, would have been hard to take. The Japanese were gratified at the comparative ease with which they drove the Russians from this fort.

On April 4 the Japanese army reached Wiju, at the mouth of the Yalu River, and found the town deserted by the 2,000 Russian troops which had occupied it. The occupation of Wiju gave the Japanese undisputed possession of the port of Yangampo at the mouth of the Yalu, which enabled them to force a passage of the river, under protection of the navy.

Japan, by this movement, compassed one of the most remarkable achievements in modern political history. She practically swept Russia clear out of the whole of Korea without anything that could be called a blow. From the northern frontier, on the wide estuary of the Yalu, to the straits looking across to the island empire itself, 600 miles away, the Hermit Kingdom was in its hands from end to end.

Never had Nemesis marched more swiftly upon the retreating track of a fatal diplomacy. Within eight weeks from the opening of the war Russia had lost the very object for which she made war.

The army concentrated at Wiju was under command of General T. Kuroki, a veteran of the Chinese war, and roughly stated, contained 55,000 infantry, 4,500 cavalry, 3,600 artillery with 180 guns, 3,000 engineers, miners and sappers, and about 3,000 men in the transport service. In all, the force comprised between 75,000 and 80,000 troops and 25,000 horses.

The actual crossing of the Yalu and the invasion of Manchuria was not effected until April 27.

CHAPTER XXVIII

THE BATTLE OF THE YALU

First Great Battle of the War Fought on Manchurian Soil—Japanese Cross the River on Pontoons under a Heavy Fire and then by an Artillery Attack Capture the Heights Held by the Russians—Flotilla of Gunboats Aid the Japanese Commander—Japanese Bodies Lay in Heaps at the River Ford—Russian Losses Heavy—Retreat Toward Fengwangcheng.

THE first great land battle of the war was fought on Sunday, May 1, at a place called Chintiencheng, not far from Wiju, but the engagement will always be popularly known as the battle of the Yalu.

The result of the battle was a brilliant victory for Japanese arms. The Russians were routed and driven back on their second line of intrenchments at Fenghuancheng. The first reports were that a force of 16,000 Japanese had surprised and defeated a force of 30,000 Russians, but the facts were that the figures were reversed. There was no surprise, for the Russian scouts were in constant touch with the Japanese forces.

Before entering upon a description of this memorable battle it will be well to follow the movements of General Kuroki's army from the time it began to cross the Yalu.

A portion of the Japanese Imperial guards attacked the Russians at Kurito island on April 26, and to enable the bridging of the Yalu to be carried out the island was occupied. A part of the second division also occupied Kinteito island and the Russians retreated towards Kiulenjo.

The Japanese casualties were nine severely and sixteen slightly wounded, belonging to the guards, and one belonging to the second division.

One man severely wounded, belonging to the Twenty-second Siberian regiment of sharpshooters, was captured. According to his statement the Twenty-third and Twenty-fourth regiments of Siberian sharpshooters were also in the vicinity under command of Maj. Gen. Tursoff. Each regiment had two battalions of infantry and 142 mounted scouts, with eight guns.

During these movements on the islands the Russians opened fire on the Japanese with eight 9½ centimeter (about 3½ inch) guns from a hill behind Chintiencheng and two Hotchkiss guns which were mounted on the banks of the river at Kosan, where the Russians seemed to have established their headquarters. The Japanese later took the Hotchkiss guns.

One battery of Japanese artillery which had taken a position on a hill to the east of Wiju fired three volleys at Kosan, where the Russian staff was located, and at noon of Tuesday the Russian batteries behind Chintiencheng shelled Wiju, wounding one Japanese soldier with shrapnel.

On Wednesday the Russians resumed the bombardment of Wiju, firing at intervals throughout the day. The Japanese artillery did not respond to this fire. General Kuroki received reports to the effect that the Russians were fortifying the heights on the right bank of the Ai river. Those defenses were declared to extend from Chintiencheng through the village of Makou to Koshoki, a distance of three and a quarter miles.

Subsequently the body of Colonel Seminoff, the head of the Twenty-second regiment of scouts, was found opposite Kurito and was interred within Wiju castle. The Japanese found ninety-five dead and six unwounded Russian horses upon the field.

The Russians resumed their bombardment on Thursday, but it was generally ineffective. Subsequently General Kuroki ordered two companies of the Imperial guards to cross the Yalu and make a reconnoissance along the left bank of the Ai.

The Japanese force advanced toward Kosan and then dispatched a small detachment to the village, where a party of Russians was encountered.

In the engagement which followed five Russians were killed. The Russians shelled the reconnoitering party from an emplacement in the hills in the southeast part of Yoshoko. This fire was without effect.

The Russian artillery on the hill behind Chintiencheng, firing at a high angle, opened on Wiju, the island of Kurito, and Seikodo, to the south of Wiju, where some Japanese batteries had taken position. This firing continued into Thursday night, and while it was ineffective, it disturbed Kuroki's preparations for an attack. The Russians resumed the shelling of Wiju on Friday, but the Japanese guns did not reply.

The twelfth division of the Japanese army was chosen to make the first crossing of the Yalu. It began its preparations on Friday by driving the Russians from their position on the bank of the river opposite Suikochin, which is eight miles above Wiju, and the point selected for the crossing. This division constructed a pontoon bridge over the river, and at 3 o'clock Saturday morning it began crossing.

The entire division passed over the river during the day, and by 6 o'clock Saturday evening it was in the position assigned to it for the battle of Sunday.

The movement of the twelfth Japanese division was covered by the Second regiment of field artillery and another artillery regiment of heavy guns.

At twenty minutes to 11 o'clock Saturday morning the Russian artillery posted to the north and to the east of Chintiencheng began shelling the patrols of Japanese infantry which had been dispatched from Kinteito island to Chukodai, another island north of Kintieto and under Chintiencheng. The Japanese batteries replied to this shelling and silenced the Russian fire. Later eight Russian guns posted on a hill to the east of the village of Makao opened upon the Imperial guards. To this shelling the Japanese artillery to the east of Wiju responded, and the Russians ceased firing.

Then both the Chintiencheng and the Makao batteries reopened, and this fire brought a vigorous response from the chain of Japanese batteries on the Korean side of the river. The Russian guns fired for two hours before they were silenced. The Japanese losses in the bombardments of Saturday were two men killed and five officers and twenty-two men wounded.

A flotilla of gunboats from the squadron of Admiral Hosoya participated in the fighting of Saturday. It encountered a mixed force of Russian infantry, cavalry and artillery on the Manchurian bank of the Yalu below Antung, and, after a sharp fight, scattered them to the hills.

A bridge across the main stream of the Yalu just above Wiju was completed at 8 o'clock Saturday night and the second Japanese division and the Imperial guards immediately began crossing. They advanced and occupied the hills back of Kosan, facing the Russian position on the right bank of the river. All through Saturday night regiment after regiment of Japanese soldiers poured across the bridge, and at a late hour Saturday night General Kuroki telegraphed to the general staff of the army:

"I will attack the enemy on May 1 at dawn."

True to his promise an artillery attack began all along the line at dawn. At 7 o'clock the Russian battery at Yoshoko was silenced and was made the point of the first attack by the twelfth division, under General Inouye.

A little stream, the Ai Ho, ran between the Japanese and Russian lines. The Japs crossed this at fifteen minutes past 8 and began storming the Russian heights. The severest fighting, around Hamatan, lasted only until 9 o'clock, when the Russians had been swept from their lines back across the plateau behind them.

Even though an attacking force numbers 3 or 4 to 1 it is desperate and difficult work to carry intrenchments, but the fighting quality of the Japanese was such that at the word of command they charged across the Ai Ho river, wading the stream breast deep, and charged up the heights in the face of a Russian fire.

The Russians were commanded by General Zassalitch, whose report of the battle was very comprehensive and is as follows:

"The Twelfth and Twenty-second regiments and the Second and Third batteries of the Sixth brigade of artillery were engaged in the battle, which began with heavy cannonading of our right flank by siege guns at Wiju and field batteries in the distance.

"After a lull the fighting was resumed with extraordinary violence against the left flank of our main position at Turenchen and our position at Potietinsky. A fusillade was also begun by small parties of Japanese across the Ai Ho river.

"The situation of the defenders' position became increasingly difficult, especially at Potietinsky, which was bombarded on the front and on both flanks.

"Thirty Japanese guns were pitted against our battery at Potietinsky, which, after having silenced the enemy's mountain battery, directed its fire on the Japanese infantry and sustained few losses so long as it was not obliged to take up another position, owing to the withdrawal of our infantry from the bank.

"The Japanese under our fire made continual attacks with fresh troops, but without having recourse to the bayonet.

"Japanese bodies lay in heaps at the river fords.

"Simultaneously with the attack at Potietinsky an attack was being made on our left flank at Turenchen and the Russian trenches had to be abandoned under the Japanese enfilading fire. Our reserves several times mingled with the first line, thus enabling it for a long time to maintain its position.

Turn Russian Flank.

"Finally all of the supports were brought up into the firing line, but owing to the great distance from our main reserves it was impossible for them to reach the advanced force in time and our men retired from the principal position to another position in the rear of Tuenchen, followed by the concentrated fire of the Japanese, who could not make up their minds to descend from the crest they occupied and face the fire of our batteries at Poulemetts. They dug fresh trenches and opened a heavy artillery fire on our new position and began to turn our left flank toward Chin-Gow.

"Two battalions of the Eleventh regiment and the Third battery of the Third brigade of artillery, belonging to the main reserve, were ordered to Lao-Fun-Hou. They occupied a position with a double firing-line, thus permitting our advanced line, which had suffered heavily, and our wounded, to retire.

"A battalion of the Eleventh regiment, both flanks of which were repeatedly turned by the enemy, advanced with fixed bayonets, preceded by buglers, to clear a passage. The Japanese, however, declined a hand-to-hand conflict and recoiled.

"In front of the regiment a chaplain bearing a cross was struck by two bullets.

"It was only by advancing on the Japanese with the bayonet that the Seventh regiment was able to retire.

"On the arrival of the battalion of the Tenth regiment all the troops were able to beat a retreat.

"The losses of the Eleventh and Twelfth regiments were very great. In the Eleventh the killed included Colonel Laming and Lieutenant Colonels Dometti and Raievski. The Twelfth lost nine company commanders, killed or wounded.

"The second and third batteries of the Sixth brigade, having lost the greater number of their men and horses, were compelled to abandon their guns after rendering them useless. For the same reason six guns of the Third battery of the Third brigade of artillery and eight poulemetts which could not be brought away, were also disabled. The mountainous nature of the country made it impossible to save the guns by means of drag ropes.

"The transportation of the wounded by hired Chinese bearers to Fengwangcheng was very difficult. Two wheeled carts and horses lent by the cavalry were also utilized for this purpose. Most of the wounded, however, arrived on foot, assisted by their comrades, and reached Fengwangcheng within twenty-four hours."

General Zassalitch's force consisted of five regiments and five batteries, but, according to the information at hand, one regiment and two batteries stationed at Shakhevze were not engaged, the Japanese not attacking that point.

The wounded priest to whom General Zassalitch refers in his

report was one of the heroes of the day. The Japanese succeeded in flanking two Russian battalions on both sides and enveloping them in the rear. The latter with music playing a martial air and with fixed bayonet, a priest holding aloft a cross, charged and broke through the Japanese lines. It was the most heroic incident of the bloody day. The priest fell among the wounded but was carried along by the escaping troops.

During the retreat a body of Russian infantry 2,000 strong occupying a hill near Tengshanghong, mistook a detachment of their own infantry, about 200 strong, which was retiring, for Japanese troops, and fought among themselves.

In the fighting which followed 110 were killed and 70 wounded, and the Russian carts were stampeded, leaving their loads of stores behind. According to the story told by a captured Russian officer, who participated in the battle, only five or six battalions of Russian infantry and two battalions of artillery were able to retire in order. The other troops ran away in a state of entire confusion.

The Russians retreated toward Fengwangcheng pursued by the Japanese. A force of the latter, however, were too eager to fall upon the retreating foe, were unexpectedly attacked and the killed and wounded numbered about 300.

CHAPTER XXIX

PORT ARTHUR CUT OFF BY LAND

Russians Abandon Fengwangcheng and Make a Stand at Kinchou—Key to the Russian Situation in Nanshan Hill which Is Strongly Fortified—Japanese Storm the Hill Nine Times in the Face of a Deadly Fire which Wipes Out Entire Companies—One of the Most Desperate Charges in the History of Warfare—Japanese Cut Railway Line to Port Arthur.

AFTER the battle of the Yalu, the first objective point of the Japanese invading army as it later developed was Kinchou, on the line of railway connecting Port Arthur with the main stem of the Trans-Siberian railway. By taking Kinchou they would effectually cut off communication with Port Arthur by land, and with Togo guarding the sea entrance the investment of the Russian naval stronghold would be complete. In addition to this Kinchou could be made an army base from which the Japanese land forces could operate southward against Port Arthur and northward against Liao Yang, General Kuropatkin's headquarters. The naval base would be Port Dalny on Talienwan Bay.

The retreating Russians fell back from the Yalu to Fengwangcheng where they were strongly intrenched. It was generally believed that General Kuropatkin, the Russian commander-in-chief, would make a stand at Fengwangcheng and endeavor to check the Japanese advance. To the great surprise of everyone, however, as soon as the Japanese vanguard came in touch with his rear guard he ordered a retreat westward. From the Yalu to Fengwangcheng there was considerable skirmishing with losses on both sides, but nothing that approximated a real battle.

General Kuropatkin had selected Kinchou as the place to make his decisive stand, and his ground was well chosen. Aside from the strong fortifications the nature of the ground favored the Russians as they were in command of all the hills and heights and were well supplied with artillery. The battle which took place there was one of the bloodiest and most desperate of the entire war and displayed the real valor of the Jap as a fighting man. The charge of the Japanese up Nanshan hill was one of the most thrilling and daring assaults ever made by an army and could only have been made by men who were without the sense of fear.

The Russians had made elaborate preparations to check the Japanese movement south on the Liao-Tong peninsula toward Port Arthur. They had fortified the high ground on the south shore of Talienwan bay, their works extending to the east and the west. The extreme Russian right was at Hushangtao and the extreme left at Nanshan hill.

This hill was the strongest part of the line. A series of batteries, strongly emplaced, crowned its crest, while rifle pits extended around its sides. Mines had been placed lower down on this hill and around the base on the northern and eastern sides were stretched well-made wire entanglements.

Another line of defenses, also protected with wire entanglements, extended from Yen-Chia-Tung, near the head of Talienwan bay, due north of Liu-Chia-Tien, which lies south of Kinchou. A strong Russian force was posted at Kinchou. It consisted of infantry and artillery.

Japs Attack in Right Angle.

The Japanese first occupied the line of hills to the east of Kinchou. Their positions had formed an almost perfect right angle, showing its southern front to Talienwan and its western front to Kinchou. Chiu-Li-Chan village was the apex of this angle. The extreme right of the Japanese lines rested at Chen-Cha-Tien, which is almost due north of Chiu-Li-Chan, while the extreme left was at Chaitsuho, a village due east of Chiu-Li-Chan. Back of this angle the attacking force assembled in complete security.

The Russians apparently attempted to draw the Japanese attack four days previous to the battle, for their batteries opened fire slowly on the enemy that day. The Japanese, however, refused

MODERN OFFICERS OF JAPAN.

WITHIN the past twenty years the Japanese Army has adopted features from all the great nations. The general officer's uniform is distinctly French; the cavalry officer wears the flat German cap.

MODERN INFANTRY OF JAPAN.

THE soldiers appear in white summer uniform, with French trousers and gaiters, and German field cap, knapsack, coat, and cartridge pouch. The sword-bayonet is both French and German.

WANDERING MUSICIANS PLAYING TO JAPANESE BEAUTY.

IN ALL the so-called holy cities of Japan are found wandering musicians, bound by vows to follow this life of minstrelsy. Their chief instruments are the flute and samisen, and they helped to pass away the time of many a Japanese beauty whose husband or lover was at the front.

to be drawn until the positions of the Russians, their guns and their strength had been fully developed.

Learn Size of Russian Cannon.

To this end the Japanese began a series of careful reconnoissances, their officers working their way close enough to the Russian position to draw the enemy's fire. They thus secured fragments of shells for the purpose of ascertaining the caliber of the Russian guns. They discovered that the batteries on Nanshan hill included four howitzers of about 15 centimeters caliber, ten old-style cannon of between 9 and 15 centimeters caliber and two quick-firing guns of 12 centimeters.

The Japanese discovered also a number of large implacements, but they did not learn the number of guns contained therein. These implacements faced to the north and to the east.

The guns fired by the Russians developed a range of 8,500 meters. Eight heavy guns posted on the Russian right in the vicinity of Hushangtao also were discovered and another strong Russian position developed by these reconnoissances was on another hill southwest of Nanshan hill, where the Russians had a series of shelter trenches.

On the shore of Talienwan bay, close to the head of the bay, the Russians had established a series of positions. Here were set up the searchlights which nightly played over the Japanese angle in the hills to the northeast.

Find Fatal Defect in Russian Line.

Further Japanese reconnoissances developed the fact that west of Liu-Chia-Tien the Russians had no defenses. Extending to the northward from Yen-Chia-Tien to the west coast of the Liao-Tong peninsula there were no defenses whatever, except the force posted at Kinchou. This gap in the defense was a fatal defect in the Russian position, and when it was perceived the Japanese extended their right to the north and east, enveloping Kinchou and the Russian extreme right.

The Japanese left also was extended to Wang-Chia-Tung, on the shore of Talienwan bay, and the center moved forward.

Wednesday morning, May 26, at 5:30 o'clock, the Japanese attacked Kinchou and for three hours they had an artillery duel with the batteries on Nanshan hill. The Russian gunners searched the Japanese lines with their fire, but failed to inflict much damage.

The battle was resumed at dawn Thursday, Japanese gunboats then entered Kinchou bay and in co-operation with the artillery on shore shelled the Russian positions.

A Russian gunboat in Talienwan bay steamed close to the shore and shelled the Japanese left. From dawn the batteries on both sides hammered away at each other.

At an early hour the Japanese infantry moved forward, and at 5:20 o'clock Thursday morning the troops entered Kinchou, the Russians retiring to the south.

The fighting continued into Thursday night, the Japanese pressing to the south and storming Nanshan hill. They followed the retreating Russians through the southern hills.

The details of the fighting at Kinchou emphasize the heroic tenacity of the Japanese in their conduct at Nanshan hill. Nine successive times the Japanese charged the fortified heights in the face of a storm of death-dealing missiles and in their last effort they carried the forts and trenches only after a bayonet-to-bayonet conflict with the Russians, who made a desperate, despairing struggle to beat back the oncoming hordes of assailants. The final assault of the Japanese, in which they succeeded in taking possession, was marked by the most desperate hand to hand encounter that has thus far characterized the war. The Japanese left throughout the entire action until night was exposed to an enfilading fire from the Russian infantry, a gunboat on Talienwan bay and four 9-centimeter guns posted at Tafengcheng.

At a critical moment the ammunition of the artillery ran low and it was decided to cast the remaining ammunition into one final desperate assault. Fortunately, however, at the moment this decision was reached the Japanese squadron in Kinchou bay, which had ceased bombarding when the infantry had first rushed forward, suddenly resumed the shelling of Nanshan hill.

Then it was that the issues of the day were determined—at a moment when the outcome was fluttering between success and defeat for Japan. At that moment was an almost certain repulse converted into victory, so successful that the forces of the czar were swept into confusion and disorderly retreat.

With every Japanese gun centering its fire upon the Nanshan forts and trenches the Japanese infantry sprang over the bodies of their dead comrades who had sacrificed their lives in the previous fruitless charges.

Rush Through the Breach.

The entire line rushed forward toward the Russian left, where the fire of the Japanese squadron had proved most deadly and which was the first to weaken under the death-dealing bombardment, and it was there that the first breach was made in the human wall that all day had been an invincible barrier to the impetuous assaults of the brown men.

It was the fourth division of the Osaka men that stormed the Russian left. It had once been said that Osaka men were not brave. It will never be said again.

The first division of Tokio, which had the center, and the third division of Nagoya, occupying the left and which had been exposed all day to the Russian fire against the front flank, now followed the example of the Osaka men and rushed forward, and the battle became transformed from an artillery duel into one of personal conflict, with the bayonet as the instrument of warfare.

Victory Crowns Last Charge.

On every parapet the restless, death-defying Japanese surged forward in increasing numbers and, hustling the Russians from their intrenchments, swept over the hill.

At 7:30 o'clock, as the sun was sinking beneath the horizon, the flag of the land of the rising sun floated above the blood-sodden

Nanshan hill, while the shouts of "Banzai!" swelled from hill to hill and re-echoed from squadron to fort.

The Japanese paid for their victory with 3,500 killed and wounded. To the Russians the humiliation of defeat was intensified by the loss of sixty-eight cannon and ten machine guns, while lying dead in the forts and trenches were 500 men, the victims of the accuracy of Japanese long-distance marksmanship and of close-range fighting.

It was in the desperate infantry charges that the Japanese sustained the bulk of their losses. In the earlier rushes of the engagement every Japanese soldier participating was shot down before he reached the first line of Russian trenches.

It was found necessary to stop these infantry charges and renew the artillery fire from the rear before the final and successful assault on the Russian position could be made. The success of this assault was brought about by one detachment of Japanese troops, more intrepid than their comrades, who succeeded in piercing the Russian line.

A stroke of fortune for the attack was the discovery and destruction by the Japanese of the electric wires leading to the mines at the eastern foot of Nanshan hill. This prevented the Russians from exploding these mines when the Japanese infantry crossed the ground where they had been placed. It is possible that the fortune of the day hinged on these mines. If the Russians had been able to explode them at the right time the losses among the Japanese troops would have been tremendous and it is possible that the Russians would have been able to hold the hill.

Nanshan was splendidly defended. Nearly fifty guns of various sizes were mounted on the various emplacements and there were also two batteries of quick-firing field pieces.

A force of Russians held Sanchilipu station, which is northwest of Dalny, but the Japanese drove them out. The Russians abandoned and burned the station and retired in the direction of Port Arthur.

The estimates of the Russians engaged in the defense of Kinchou, Nanshan hill and the south shore of Talienwan bay vary, but it is evident that the Russians drew men from the forces at Port Arthur and offered all the resistance possible.

It is understood that Lieutenant General Stoessel, commander of the military forces at Port Arthur, was in personal command of the Russian operations at Kinchou.

The Japanese squadron, which was assigned to co-operate in the attack of the second army on Kinchou and Nanshan hill, was composed of the gunboats Thukishi, Abagi, Heiyen and Chokai and the first torpedo-boat flotilla. It entered Kinchou bay Wednesday. A heavy sea prevented its participation in the fighting of that day. The weather cleared Thursday morning, however, and in spite of the shallowness of the water the squadron steamed close to shore and bombarded the Russian batteries.

Early in the action a shell passed over the forward deck of the Chokai, killing a lieutenant and two petty officers and wounding two men.

The torpedo-boat flotilla shelled the railroad near the Shaos river. The torpedo-boats took soundings and guided the larger ships. The squadron advanced with the Japanese right flank and aided in covering it. Later in the day the falling tide compelled the withdrawal of the larger warships.

Captain Hayashi, commander of the Chokai, was killed by the explosion of a shell near one of the ship's guns. Four other men were wounded. The vessels themselves were not damaged.

The Japanese in attacking Kinchou and Nanshan hill had to fight against great odds. The Russians were in full command of the strategical advantages afforded by nature and these advantages were augmented by the newest inventions for defense. The forts on Nanshan hill were armed with heavy guns. The Japanese had only field guns, heavy guns being unavailable on account of the difficulties of transportation.

The Japanese army deserved great credit for having driven the Russians from this stronghold; it was a feat previously considered to have been impossible.

By their great battle and the capture of Kinchou and the forts around the city the Japanese opened a free passageway to Dalny and Port Arthur, though at the cost of thousands of soldiers, for the battle was most deadly for the Japanese as well as for the Russians.

The battle of Kinchou marked the actual beginning of the siege of Port Arthur, and enabled the Japanese to concentrate all the land forces not needed for the siege against the main Russian army. Thus it was that while Nogi with a besieging army fought his way south, taking town after town and fort after fort until his men came into hand-to-hand conflict with the Russians in the last fort at Port Arthur, three armies under Kuroki, Oku and Nodzu fought the brilliant series of victories ending with the drawn battle at the Sha river.

Kinchou was the Gordian knot, so to speak, that bound Port Arthur to Russia, and when it was cut the Russian campaign received its most vital blow. The capture of the commercial city of Dalny, which followed soon after, gave the Japanese a most desirable port and greatly aided their transport service both in men and supplies.

CHAPTER XXX

DEFEAT OF STAKELBERG'S RELIEF EXPEDITION

General Kuropatkin Sends an Army Corps under General Stakelberg to the Relief of Port Arthur—Description of the Battles of Vafangow and Telissu in which the Russian Forces Are Cut to Pieces and Flee Northward to Rejoin the Main Army Under General Kuropatkin—Japanese Are Left Free to Attack Port Arthur by Land.

THREE days after the battle of Kinchou and Nanshan hill, which cut off Port Arthur, the Japanese occupied the commercial city and port of Dalny, which gave them a naval basis from which to land fresh troops and supplies for the victorious armies.

The Japanese were now in a position to march on Port Arthur and assault it from the rear, while Admiral Togo attacked by sea. The critical position of Port Arthur was such that General Kuropatkin dispatched an entire Russian corps under command of Lieutenant General Baron Stakelberg to the relief of the besieged fortress. In a battle which began at Vafangow, about fifty-five miles north of Port Arthur, and ended at Telissu, about twenty miles still further north, General Stakelberg's relief column was enveloped, surrounded, cut to pieces and utterly routed by two Japanese columns under Generals Nodzu and Oku, respectively. The defeated army fled to the northward toward Kuropatkin's base at Liao-Yang on the road to Mukden.

The Vafangow operations really began on June 11, when two Japanese divisions, one commanded by General Nodzu, and one by General Oku, advanced from the Pulantien-Pitsewo line, prompted by intelligence that the Russians were receiving reinforcements. A sharp skirmish took place the night of June 12 at the village of

Oudiaden, five miles west of the railroad and seven miles south of Wa-Fang-Tien. Another skirmish occurred on the heights near the village of Lidiatun, ten miles east of Oudiaden. The Japanese were repulsed, but the Russian advance posts retired. June 13 two Japanese divisions were five miles north of the scene of the skirmish, their left wing resting on the village of Vafangow and their right on the valley of the Tassa, which flows parallel with the railroad, falling into the sea ten miles south of Pitsewo.

After a day's rest the Japanese advanced fifteen miles June 14 and attacked the left of the Russian position, four miles south of Vafangow. The Russians held a line between Lung-Wang-Tiao and Ta-Fang-Shen. The Japanese artillery opened on this line and the Russians responded. The shelling continued for two hours and it was followed by the advance of the Japanese line to a position extending from Lung-Chia-Tung to Yu-Hotun.

Darkness put an end to the fighting. The Japanese dispatched a column to the westward toward Fuchau for the purpose of covering the Russian right wing and to protect their left and rear. During the night it became apparent that the Russians were being reinforced and it was decided to make a general attack in the morning and force the Russians into a defile back of Telissu.

When morning came it was discovered that the Russians held a line extending from Ta-Fang-Shen to Cheng-Tsu-Shan.

The Japanese planned to envelop the Russians near Telissu and they succeeded admirably. While the main Japanese force was moving north along the railroad, columns swung to the left and to the right and finally converged at noon on the main Russian position. The Russians in this position were at a disadvantage, but they held it with determination until 3 o'clock in the afternoon. At this hour they were routed.

Artillery, cavalry and infantry were engaged. For the first time in large numbers the Japanese cavalry clashed with the famous and greatly dreaded Cossacks and rendered a good account of themselves.

The Russian reports of the battles of Vafangow and Telissu

are highly colored, but it should be stated in all fairness that the Russian troops fought with valor and were bravely officered. It must be remembered that the Japanese largely outnumbered the Russian force, but it is equally apparent that Nodzu and Oku outgeneralled Stakelberg at every point. They fooled him repeatedly by maneuvering their troops and finally forced him into a trap at Telissu exactly as they had planned. The stern, dogged fighting at the battle of Vafangow was like another Borodino. Throughout the three days of combat the Russian officers and men vied with each other in pluck and heroism.

The Japanese advance originally included the Fifth, Eighth and Eleventh divisions, twelve squadrons of cavalry and splendid artillery. About 200 guns were belching a continuous stream of shot and shell by which they were enabled to turn the Russian flanks. A diversion on the right precipitated the battle on the morning of June 15.

Major General Gerngross, who was wounded, commanded the Russian left flank, and General Loutchkovsky commanded the center, including four battalions concealed in a small wood, whence they dealt death and destruction on the attacking forces. The Russian right was protected by Cossacks, dragoons and Siberian rifles.

The scene was awe-inspiring. Over the Russian center and left flank hovered chocolate clouds from bursting shrapnel. It was evident about noon that the tide of battle was turning toward the Russian right. Reserves hurried forward, the Cossacks galloping, followed by columns of infantry at the double. Suddenly they disappeared in an adjacent defile. The valley where the Russians had camped was emptied as if by magic. Rattling volleys were fired behind the screen of hills which concealed the fighting troops from view in that direction. This continued for half an hour. Suddenly a company of Cossacks appeared on the crest of a hill and began to descend. They were followed by infantry. The Japanese gunners promptly pursued them with shrapnel. Horses and men began falling.

A moment of harrowing suspense was relieved by a thunderous shout of "Hurrah!" It was from a couple of thousand of Russian troops just brought up by train. They quickly jumped from the cars, fixed bayonets and literally ran into the fight.

Again the crackle of musketry under cover during which the Russian lines broke and fled toward the railroad. While a long line of commissariat wagons, escorted by Cossacks, took to the road, a battery of horse artillery stationed near the railroad banged away furiously as it covered the retreat. The Japanese shells were falling on the station buildings as the Russians hurriedly entrained.

Neither the Russian nor the Japanese commanders made any attempt to conceal their losses in this battle, which up to that time was the most disastrous of the war. In a report to General Kuropatkin the Russian losses were admitted to be heavy, although at that time they were not known. General Stakelberg, in his report, dated June 16, said:

"Yesterday I had intended to attack the enemy's right flank, but just as our troops had been assigned for the purpose and were beginning successfully to envelop the enemy's right flank the Japanese in their turn attacked my right flank with superior forces, and I was compelled to retreat by three roads to the north.

"Our losses are heavy, but they are not yet completely known. During the engagement the Third and Fourth batteries of the First artillery brigade were literally cut to pieces by the Japanese shells. Of sixteen guns thirteen were rendered completely useless and were abandoned.

"The conduct of the troops was excellent, a large proportion of them refusing to retire until after they had been repeatedly ordered to do so."

The fierce character of the fight is made evident by the fact that the Russians were again forced to abandon their guns, thus indicating, as in previous encounters, the superiority of the Japanese artillery.

General Oku reported as follows;

"Our main body advanced northward June 14 in two columns along the railway, expelling the enemy from the east of Wafangtien. At 5 p. m. the enemy made a stand on the railway between Lungwangmiao and Tafangshen and after two hours of cannonading we occupied at nightfall the line between Pangchiaton and Yuhoton.

"Another column, for menacing the enemy's right and covering our flank and rear, marched eastward of Foochow and advanced on the line between Tengchiakow and Nachialing. The enemy near Lungwangmiao was reinforced.

"June 15 the enemy near Telissu consisted of two and a half divisions, occupying the position between Tafangshen and Chengtsushan. At dawn we opened attack and our main body advanced along the railway, one column marching from Tsouchiaton. At 9 a. m. the left wing of the column was joined by the forces from Tungtungkow and at noon by cavalry from Chiachiaton. Thus the enemy was surrounded by our forces near Telissu and after severe fighting they were routed and fled northward at 3 p. m.

"Our losses are estimated under 1,000. So far as known we captured colors, fourteen quick-firers and about 300 of the enemy, including the commander of the Fourth infantry regiment of sharpshooters.

"Over 500 of the enemy's killed and wounded were left on the battlefield.

"Our scouts saw the enemy marching with the Japanese flag in this engagement, by which our artillery was misled and suspended fire."

The estimated losses were: Russians, killed, 1,854; wounded, 3,500; captured, 300. Japanese, killed, 247; wounded, 946.

General Stakelberg's army consisted of the first, second and sixth East Siberian rifles divisions, the Usuri mounted brigade and the first East Siberian sapper battalion.

General Gerngross, who was wounded, commanded the first Siberian division of four regiments and a machine gun battalion.

General Samsonoff commanded the Usuri mounted brigade.

A lieutenant of Cossacks says the American military attaches were with his command most of June 15, during the hottest part of the fight. He commented upon their coolness and their professional interest in the operations to the exclusion of the idea of personal danger.

"The Russian artillery," the lieutenant says, "was splendidly served, but was outmatched in number by the Japanese guns. One Russian battery pitted against Japanese batteries was literally smothered by Japanese shells. I saw one Russian battery land three shells in the midst of an ammunition train which was galloping up to serve the Japanese guns. Two caissons exploded, killing all the horses and drivers.

"The Japanese guns fired at least 1,500 rounds. The Russians fired several times on Japanese infantry in close formation, causing tremendous havoc."

A Russian officer of the fourth battery says his battery was in a duel at two and a half miles with a Japanese battery and silenced it. His battery then ran out of ammunition and the men carried the breech mechanism of the guns with them to the rear and brought up a fresh supply of ammunition and resumed serving the guns.

A war correspondent who was present at the battle on the 15th and 16th says the work of the Red Cross in attending to the wounded under fire was beyond all praise. One Red Cross assistant was killed and another was wounded.

The Tobolsk regiment saved the last hospital train from Vafangow by gallantly throwing itself between the train and a superior force of Japanese and by holding off the enemy while the train steamed out under a heavy fire.

Shell fire burned the Red Cross station at Vafangow, destroying almost all the supplies belonging to the Sisters of Mercy and much hospital material.

There seems to be good ground for the belief that the Russians saved themselves from heavier losses at the beginning of the retreat by hoisting a captured Japanese flag, and thus causing some of the Japanese batteries to cease firing.

The battle of Vafangow and Talissu was a defeat for the Russians in more ways than one. Not only was the crack corps of Siberian rifles badly cut up and routed, but the purpose of the expedition which was to draw off a part of the force mobilizing against Port Arthur was likewise defeated and no material aid was rendered to the beleaguered fortress.

From that time on General Kuropatkin was never in a position to again attempt to relieve the besieged fortress. With three Japanese armies pressing upon his front his entire attention was given to his own defense.

In the meantime General Stoessel, completely cut off from any chance of relief by land, continued to defend Port Arthur, entirely ignorant of the situation north and therefore hoping and believing that Kuropatkin was coming to his aid.

Indeed, as it afterward appeared, Stakelberg's corps would have been of little assistance to the beleaguered fortress even if he had succeeded in getting through the Japanese lines, for the superior forces of the Japanese would eventually have crushed the life of the small expedition.

Although General Stakelberg's command was defeated and badly cut off, the commanding officer deserves great credit both for strategy and bravery. When he was practically surrounded and his command on the point of being captured, he fought and maneuvered with such skill that he was able to rejoin the main army, and thus save the remnant of his corps.

CHAPTER XXXI

THE "THERMOPYLAE OF MANCHURIA"

Kuroki Defeats Kuropatkin at the Second Battle of Motien Pass and Opens the Way to Liaoyang—Russians Lose the "Key to Manchuria" after a Desperate and Bloody Battle Fought in a Fog—Japanese Occupy the Valley of the Liao River—Personal Experiences of a War Correspondent.

MOTIEN PASS has been called the "Thermopylae of Manchuria," and was considered to be the key to the whole situation so far as the land forces were concerned.

Admirably situated for defense in the Fenshui mountain range, it was supposed that a small force could hold it against a large body of invaders. The battle which took place there was important, not from the number of men engaged, but from the fact that it enabled the Japanese to occupy the valley of the Liao river and later to rout the Russians at the battle of Liaoyang.

When the war began everybody in Tokio believed that the Russians were going to make a great stand at the famous pass. As a matter of fact, the Russians let the Japanese seize the pass without much trouble in June. Until the Japanese had reached the pass they had been going up-hill. From that to Liaoyang (fifty-five miles distant), where they were to fight to such good purpose, the road is down-hill.

After some deliberation the Russians resolved to reoccupy the pass, seeming to realize when it was too late its strategical importance and its value as a defensive position.

First Fight at Motien Pass.

The first attempt of the Russians to recapture the pass was on July 4, when three battalions of infantry attempted a frontal attack, thrice repeated. They were driven back without difficulty, losing 200 men, although the pass was held by but one Japanese battalion. Three days later the First Regiment of Cossacks, numbering 1,300 sabres, advanced against Fen-shui-ling defile, but without pressing home their assault they retired, neither inflicting nor receiving much loss.

The second attempt to recover the pass was made on July 17, under Lieut.-General Count Feodor Keller (who was killed at Haicheng on August 1). The Russians had 20,000 men, including the 9th and 24th East Siberian regiments and the 34th Regulars of the 9th Division, the first Russian regulars to meet the Japanese.

Fighting in a Fog.

The first intimation to the Japanese of the Russian movement was obtained half an hour after midnight (July 17) from General Matsanaga, who was in command of a brigade on the Japanese left front. He reported that a party of Russians was moving on his front and warned the commander of the force at Motien pass to be on his guard. General Okasaki immediately turned out his brigade and waited in readiness for the development of events. The first sign of the enemy near Motien pass reached the Japanese at half past 2, when 150 cavalry and a large body of infantry appeared close to the line of Japanese pickets. The morning was so misty that the Japanese could not see a hundred yards in front of them. The Japanese defensive line was along the ridge at the top of the pass, and in view of the vast superiority of the enemy the outposts slowly retired in that direction.

About 6 o'clock the fog began to rise from the hollows, disclosing the main Russian force lying in the valley directly opposite the lofty flanking ridge which dominates the approaches to the pass. Upon this ridge 2,000 yards away from the Russians was the Japanese artillery position so far unoccupied by guns. The Japanese climbed the high ridge on the left of the pass and from various points poured a hot enfilading fire down upon that part of the

Russian line nearest the pass (it was within 400 yards) and forced it to withdraw into an extensive wood immediately in rear.

At a quarter to 7 the Japanese guns, after great labor, gained their position and opened a devastating fire upon the Russians massed in the valley before them. About 7 o'clock the Russians had reached the limit of their advance. They occupied the temple, driving out at the point of the bayonet a small party of Japanese who had been left there to delay their advance. The Russians scaled the low boundary wall of the building at the rear and effected an entrance in that way. They afterwards used the wall as cover from which to fire at the Japanese trenches across the valley. Nearly a score of dead Russians were lying round the temple. The main body withdrew to shelter, leaving hundreds of dead and wounded in their hurried retreat. When the temple was evacuated the remainder of the Russians fell back upon the wood.

The Japanese guns ceased about 8 o'clock, by which hour the Russians on the right were out of action. The fighting was confined to the left.

To the astonishment of everybody, about 9 o'clock the Russians began to retreat. This was attributed to a bit of "bluff," some reinforcements having come to the Japanese. The Russians retired with the utmost deliberation.

After the battle the air was full of summer scent, and the bushes were dotted with gorgeous flowers of the crowning days of a Manchurian July. A bright sun was beating down pitilessly, for the fog was gone. The bushes were marked with stiffening corpses, blue-eyed, light-haired, and often with eyes wide open in death.

The Russian casualties amounted to 2,000. The Japanese had forty-three killed and 256 wounded. Among the Japanese killed was Major Hiraoka, who was the Mikado's military attache during the South African War.

No war correspondent had a better opportunity to view the engagement at Motien pass than E. F. Knight, who was with the

HUMAN BIRDS OF PREY.

Horrors of the Battlefield Intensified by Chinese Bandits, Robbing the Dead and the Dying.

RUSSIAN ARTILLERY ON THE MARCH IN MANCHURIA.
The Bitter Cold and Heavy Snows Make Progress Very Slow.

central force, known as the first army, under the personal command of General Kuroki. He has supplied the following record of personal experience:

"The Japanese positions—I am speaking now of the central of the three forces composing the First Army under General Kuroki's command—extended along the ridge of the range which is crossed by the Mo-tien-ling and the main road to Liaoyang. Yesterday morning, the 16th, shortly before 1 o'clock, the Japanese received intelligence that the enemy was moving on their front, and an hour and a half later a large body of Russian infantry, with some cavalry, came in touch with the Japanese line of pickets, which then fell back, in accordance with orders, on the line of defenses above. At dawn the fighting commenced, for on these hills the mist was not so dense as we had found it in our valley camp seven miles to the rear. The fog gradually lifted, and the broadening light disclosed to view the main body of the enemy advancing up the broad valley along which the road to Liaoyang is carried. The Russians were in close formation, so that from the heights the Japanese guns and infantry were able to direct on them shrapnel fire and rifle volleys with deadly effect. Those who traversed this portion of the valley later in the day found the ground strewn with Russian dead, and the burial of these is still being continued by the Japanese as I write this dispatch.

"The main attack was made on the pass itself. But the fighting was by no means confined to its immediate neighborhood. The Russian attack was directed at several points along the Japanese strategical position, and extended along a front of fifteen miles. Nearly two Russian divisions—seven regiments according to reliable information—were engaged, so that the enemy probably numbered about 22,000 men, possibly more, for the full strength of a Russian regiment is nearly 4,000 men. On the Yalu the Japanese were opposed by regiments of the East Siberian army. But the Russian troops that fought in the Mo-tien-ling mostly belonged to the Ninth Army Corps from Europe, and were under the command of General Keller.

The Enemy's Object.

"It was apparently the enemy's object to turn the Mo-tien-ling from the north. For it was on the Japanese right that the attack was driven furthest home and that the most severe fighting occurred. At this point, which the Russians attacked with three regiments, the Japanese suffered their heaviest losses. Here the enemy attained the summit of the ridge before they were driven back, and it is difficult to understand how it is that, being in such greatly superior force, they did not retain the position they had gained. Here the Japanese must have offered a most gallant resistance. For example, an outpost composed of one company was surrounded on three sides by six companies of the enemy, which approached within one hundred yards; the Japanese company lost twenty killed and fifty-four wounded, including all the officers and sergeants, but the men carried on their resolute defense until a portion of their regiment came to their relief, when the enemy retired. Another company was for a time engaged with an entire Russian battalion, and a sergeant's post of obseravtion of twelve men successfully held its own behind trenches against a Russian company. The fighting was also for some time very severe at the Mo-tien-ling, where the enemy's main attack was vigorously opposed. The Russians fell back in very good order, holding for some time the heights that flank the Liaoyang road in order to cover their retirement.

"On the Japanese left there was little fighting. It was on this point that the Russians first advanced shortly after midnight, but with three companies only, so that this early move was probably intended as a diversion to cover the real attack on the right. A brigade on the Japanese left made a demonstration against the main body of the enemy.

"Of the three forces composing General Kuroki's army the central one alone, therefore, was the object of the Russian attack this day. Far away, on the right of this force, extended the defensive positions of a second force which took no part in the action; while beyond our left wing was stationed the third force, which also was not engaged, with the exception of one battalion with some guns, which attacked the enemy's right flank and shelled them during their retirement. I have advisedly used the word 'force,' as no term which might convey an idea of the strength of General Kuroki's army would pass the censorship.

"When the enemy's retirement became general the fight slackened and became what one might almost describe as a leisurely one, the enemy retiring very slowly, the Japanese not pressing them hard. But it was not until dusk that the last shot was fired. In the night the enemy fell back on the positions from which they had advanced in the morning. The Japanese casualties were 341, out of whom forty-three only were killed. The Russian losses are unknown. The enemy carried away large numbers of their dead and wounded, and are said to have buried 500 at one place. Their dead were lying thick in the valley below the pass. It is estimated that the Russian killed and wounded must have numbered about 2,000.

Personal Experiences.

"And now to say something of what I saw of the fight when I reached the western slopes of the Mo-tien-ling, where the enemy's main attack was made on the Japanese center.

"Magnificent was the scene that spread before us. The wooded, and in some parts grass-grown, slopes of the Mo-tien-ling fell from our feet down to the broad valley which stretched out straight in front of us in the direction of Liaoyang, and we saw winding down the valley a tributary of the Tai-tse, the river that flows past the city of Liaoyang. The slopes of the Mo-tien-ling are broken into many swelling spurs, and down the converging gullies rush the little streams that unite to form the river below. We descended the slopes of the pass, up which stretcher-bearers with wounded were slowly climbing, and soon we began to realize that the Russians had driven their attack a considerable way up in this direction, for we saw the bodies of Russian dead lying on the road, where there were also numbers of blanket-rolls, provisions of black bread, water-bottles, and so forth, which the Russians had thrown away when retreating.

Care of Enemy's Wounded.

"In many parts of the field we came on the Russian wounded remaining under the guard of Japanese soldiers until the stretcher-bearers should arrive. I see from quotations appearing in the Japanese papers that the atrocity-mongers are already at work with their malicious inventions. There are unreasoning sentimentalists in Europe who give ready credence to these tales, however prejudiced or irresponsible their origin, all the more so if the supposed victims belong to a nation professing Christianity, whereas the alleged perpetrators of the barbarities do not. At the present moment there are at the front with the Japanese forces about a score of European military attaches and newspaper correspondents, and I am sure that all of these will testify that so far as their own observation goes the Japanese treat the prisoners and wounded of the enemy with great kindness. It will be suggested, perhaps, that the Japanese carefully keep us out of sight of their misdeeds. But this would be impossible. We are confined by no very limited bounds here, as was the case at Feng-whang-cheng. We ride and walk freely and unescorted along these roads at the front, and we are ever coming quite unexpectedly on Russian prisoners and their captors. The Japanese could have had no time to arrange a picture for us. So, too, was it in the heat of the action this day on the Mo-tien-ling: whenever we came on a wounded Russian in the hands of Japanese privates they were seeing to his comfort, cheering him up, binding his wounds, offering him cigarettes, and, in short, proving themselves most humane foemen. Often at Antung, at Feng-whang-cheng, and here we have conversed with Russian prisoners, and invariably have they told us—generally volunteering the statement—that they are grateful to the Japanese for their kindly care of those who fall into their hands. Not one exception to the rule has yet come to our notice. Such is the evidence that we who are at the front can give.

Leisurely Retreat.

"So as to obtain a better and closer view of what was doing I and others ascended a steep conical hill on whose summit the Japanese had a post of observation. From here I saw that a large body of Russian infantry was massed in front of a wood on a broad flat space about three miles up the valley, and that between this force and us other bodies of the enemy were retiring slowly along the tops of the hills flanking the valley, firing volleys at the Japanese, who from the nearer hills and from the flats below us were opening fire on the different groups of the enemy. Other parties of Russians were quite leisurely retiring up the valley towards the massed force under the wood. A number of Russian ambulance wagons were also carrying the wounded in the same direction.

"For some time the retirement seemed to cease altogether, the Russians holding their ground; and, as I have said, it was not until after dark that they fell back on the positions from which they had advanced in the morning. For several hours we wandered about, observing a scene that in some way reminded one more of maneuvers than of actual warfare, but the dead bodies around us brought one back to a recognition of the stern reality of the business. While we were on the field the Japanese brought no guns to bear on the enemy. Neither did the Russians bring any guns into action until about 3:30 p. m., when one of their batteries directed an accurate shell fire on a near spur immediately facing us, causing the Japanese who lined it to retire under cover of the ridge. The Russians covered their retreat well, and had the Japanese pressed the pursuit of this force, so superior in numbers to their own, their losses would have been very heavy and no adequate advantage would have been gained. The firing gradually slackened, and at 4 o'clock, as the engagement was practically over, we rode back to camp.

"As I have before pointed out, in view of the strength of the force they employed, the retirement of the Russians is difficult to explain if it was really their object to capture and hold the pass. Their retirement was slow and orderly, as of men who have accomplished the work assigned to them, and are leisurely returning. Was this, then, a reconnaissance in force that had orders to retire when its purpose was effected? Was it a demonstration to mask a battle in the west, or to cover the beginning of the retreat of Kuropatkin's army? Whatever was intended the Russians lost heavily."

Following the battle known as Motien pass, the second Japanese army under General Nodzu attacked the command of General Zassalitch, intrenched at Haicheng, on the main road to Liaoyang and Mukden, and after two days of severe fighting captured that important situation and drove the Russians back to Anshanshan. The battle was fought in terrible heat and many of the troops in both armies suffered severely, while others succumbed to sunstrokes. In this battle General Feodor Keller, the first Russian general to be killed in the war, lost his life by the explosion of a shell which burst within a few feet of him.

General Zassalitch later made the claim that he was on the point of winning a victory when he received an order to retreat.

CHAPTER XXXII

THE BATTLE OF LIAOYANG

Six Days' Battle between the Russians under Kuropatkin and Three Japanese Armies Ranks as One of the Great Engagements of the World and the Bloodiest Since the Franco-German War—Losses on Both Sides Almost as Great as at Gettysburg—Defeat of the Russians, Who Were Strongly Intrenched, Assures Japanese Control of Manchuria.

LIAOYANG will rank in history as one of the great battles of the world. In some respects it was not unlike the fiercest battles of the American civil war. Although not a Sedan, it was the bloodiest encounter since the Franco-German war.

In any case the battle will be known in history as the first pitched battle of the highest importance which has ever been fought between the East and the West. Liaoyang was regarded by the Russians as a point entirely in their favor in the adroit series of retreats which marked Kuropatkin's tactics (as influenced by the desire of the officials in St. Petersburg to make a stand). He had fortified the place very strongly under the supervision of a great engineer, General Velitchko, at a cost of 23,000,000 roubles, and yet the Japanese drove the Russians out.

The battle proper began on the evening of August 29 and lasted for six horrible days.

General Kuropatkin faced three armies, namely, those of Oku, Nodzu, and Kuroki. The fight began by a bombardment of the Russian positions on the Shoushan range, which lies some four to six miles south of Liaoyang, but the day was inconclusive except that Kuroki managed to cross the Taitse river at a point ten miles east of Liaoyang, and thus threatened the Russian flank. On the third day the Japanese delivered a successful attack on the southwest of Liaoyang and drove in the Russian right wing and then Kuropatkin began another of his famous retreats to the north across the Taitse river. The Japanese immediately started in pursuit and turned some of the Russian guns on to Liaoyang station.

At a quarter to 4 on Friday afternoon, September 2, the first Japanese shell fell into Liaoyang and was followed by a hail of projectiles which swept the railway station, the suburbs, and the town itself. Luckily for the Russians the railway station was empty, all the rolling stock having been removed. The first persons wounded were a sister of charity, a doctor, and several Chinese, as well as a non-commissioned officer of the transport service. After this the Japanese had to fight practically a rear-guard action.

The troops had to fight without food for two days. The Russians fought valiantly as they always do when driven to bay, but they were completely outgeneralled by Japanese brains.

Liaoyang, to which the Russians retreated on August 2, after a disastrous defeat at Haicheng, lies in the rich valley of the Liao river. Spurs of the mountain chain which forms the backbone of Manchuria slope down westward nearly to the town, and a ridge cuts the plain southward, through which the Port Arthur-Harbin railway runs. The town is about forty miles from Tashichiao, also the scene of a Russian defeat, which was responsible for the loss of Newchang, and about thirty miles from Mukden, used by the Russians as a base for the disposition of troops.

Months of labor had been expended on the defenses of Liaoyang. The fortifications and intrenchments, planned by General Velitchke, were popularly believed to be impregnable. The lines ran in a semi-circle along the ridges from eight to ten miles east and south of the town. Galleries protected the artillerymen, and wire fences and mines were disposed to repel infantry assaults. All parts of the works were connected by telephone, and were considered to be marvels of engineering skill. Within these lines General Kuropatkin withdrew an oftimes beaten, but plucky, force of infantry, cavalry and artillery. Estimates of its strength,

UNDER FIRE.
Japanese Artillery Crossing a River.

THE MARCH OF DESPAIR.
Wounded Russians on the Way to Liaoyang.

THE COSSACK GENERAL, RENNENKAMPF,
In the Rocky Fastnesses of Manchuria.

SELLING LIFE DEARLY.

A Wounded Japanese Lieutenant of the Second Infantry, with Four Comrades Down, Stands at Bay, and Kills Several More Russians before Death Comes.

with reinforcements from the north, varied from 170,000 to 225,000 men and from 700 to 1,000 guns, many of them emplaced.

The relative position of the Japanese armies was practically unchanged when operations against General Kuropatkin were resumed near the close of the rainy season, on August 23. The three Japanese forces covered all the points of the Russian semicircle—Kuroki on the east, Nodzu on the southeast and Oku on the south. Their total strength was estimated at 240,000 men and 1,000 guns.

Field Marshal Oyama, the Japanese commander, set the combined armies in motion before the rains had ceased and while transportation and movements were difficult. The Japanese began a series of sharp attacks on the outlying Russian position on August 24, and on the following day two divisions of Kuroki's left were in readiness to attack Liandiansian, about twenty miles southeast of Liaoyang.

This attack was delivered early in the morning, and lasted two days, while Oku and Nodzu, commanding the Japanese left and center, respectively, assailed the Russians on the south and southeast. Nodzu's left was beaten back and Oku's direct success was small, but Kuroki rolled up the Russian left, inflicting heavy losses, and forced the retirement of all the Russian corps on the main defenses. Anshanshan, their strongest position on the south, was evacuated under heavy pressure from Oku, and several guns fell into his hands.

In the beginning of the battle, after General Kuroki had noted the march of the Tenth Russian army corps, which he had driven in from the Anping road to the plain, on its way to join the Russian center, which General Nodzu had forced in from Kaofengshi, he divided his forces stationed in Anping and Fengwangcheng roads. He took care to have a sufficiently strong force on the Fengwangcheng road, on General Nodzu's extreme right, and sent to General Nodzu, who was pushing with the Japanese center army along the Mengchiafang and south roads, one brigade to Mengchiafang, three regiments to Chiaofantun and two regiments to the south road. General Kuroki, in the meantime, waited quietly with two divisions on the south bank of the Taitse river.

Oku Advances Under Fire.

On August 29 the artillery under General Nodzu bombarded fiercely the three hills at Chiaofantun. On the same day General Oku, in command of the Japanese left army, working his way with rifle fire and artillery, slowly advanced along the railroad in spite of the Russian opposition from neighboring elevations and fortifications. The left army thus moved toward Shusean. The Russian cavalry was not able to locate the Japanese trenches on this advance.

The character of the Japanese advance apparently had served as a warning to General Kuropatkin to make sure of his retreat and his transport was at once seen moving rapidly out of Liaoyang over the railroad bridge.

On August 30 the Japanese closed in on the Russians in their mountain position at Chiaofantun. Here the guns were so close to each other that it was almost possible for the battery commanders to see each other's spectacles.

Unable to Locate Japs.

At 10 o'clock on the morning of the 30th General Kuropatkin's staff moved out of Liaoyang. At the same time rifle fire was heard drawing nearer to the railroad station, and this was an indication of the sure advance of the Japanese left army.

The fences around all the houses at Liaoyang had been removed in order to facilitate the movement of wagons and troops. Wounded men, of whom none was seen on the 29th, were trailing along the roads from the south and southwest into Liaoyang on the 30th. The Russians made use of a field balloon all day long on the 30th, and by means of it they got the range of one Japanese battery, which they succeeded in putting out of action. As a general thing, however, they failed to find either the enemy's batteries or his troops, while the Japanese continued to silence Russian battery after Russian battery.

Russians Become Stronger.

At 3 o'clock in the afternoon of the 30th the Japanese shelled a Russian battery posted on a hill within three miles of Liaoyang, but shortly after this hour it became apparent that the Russians had begun to hold their own. By 4 o'clock in the afternoon the Japanese shrapnel fire was confined to the position at Chiaofantun, where the Russians were disputing every hillock with the enemy.

The Japanese made repeated but unsuccessful assaults upon this position. The fighting of the 30th came to a close at 7 o'clock in the evening. The Russian casualties for the 29th and 30th totaled more than 5,000 men.

The Japanese, from a low range of hills to the southeast, and south of Shushan, began shelling the northern end of the Russian line at Shushan at 9:45 o'clock on the morning of the 31st. This artillery fire was preceded and accompanied by infantry rushes over the valley and the lower hills south of the road.

Urge Men With Swords.

They began before dawn and were continued until evening, the Japanese officers urging their wornout and fainting men with the points of their swords. There was very hard fighting at a round-topped hill in the Russian line opposite the point where the two Japanese armies under Generals Nodzu and Oku came together. The Russian position here was protected with wire entanglements and a small body of Russian troops absolutely refused to retire.

In one trench seven men fought gallantly until they were surrounded by the Japanese. At the end of this sharp fighting the Russians were hurling down sandbags upon the enemy. They succeeded in killing one Japanese officer and injuring another, after which they surrendered and the hill was occupied by the enemy.

At 5 o'clock in the evening the fighting, which had raged all day long, centered along the south road, where for two hours there was a heavy artillery fire and a vast amount of ammunition was expended. The valley in the front and under the Russian guns during this artillery duel was an inferno of shrieking and moaning shrapnel, while beyond the Russian shells burst continuously at the head of the Japanese advance.

During the day the Russian rifle fire was indescribably furious. The Russian trenches, for a distance of several miles, threw out all day long a continuous hail of lead, accompanied by unceasing flashes of fire. In spite of its continued ferocity this rifle fire was comparatively ineffectual, the Russians often firing without sight of the enemy. This hail of Russian rifle bullets did not cease before 10 o'clock at night. It is estimated that 1,000 Russian shells were thrown into the valley beyond the south road, where the only evidence of the presence of the Japanese was a Red Cross camp.

Russians Start to Fall Back.

At nightfall General Stakelberg, with the first corps, was still facing General Oku on the left. This Russian corps had lost 3,000 men.

During the night of the 31st the Russians deemed the position of the enemy to be so threatening that they made preparations to fall back.

General Mistchenko was on the supreme right of the Russian line with a division composed of cavalry, artillery and infantry. During the night of the 31st his force was sent out in a northeasterly direction to gain information concerning General Kuroki's movement around the Russian's left flank in the direction of the railroads. General Mistchenko's movement was of short duration and he soon returned to his line.

General Ivanoff and his command was stationed on General Stakelberg's right and facing General Nodzu and the Japanese center army. General Ivanoff had succeeded the late Lieutenant General Count Keller to the command of the Russian eastern army. He had moved his headquarters from the village of Tashi, in the middle of the southern plain, to a point within the shelter of the city walls and on the road which leads to Yentai by a pontoon bridge across the Taitse river.

Transports Move from City.

During the night of the 31st the Tenth corps, which had assisted in holding the Fengwangcheng road against the Japanese, began moving its transport out of the city to the north.

The Seventeenth corps, which had not as yet been in action and which was being held on the north bank of the Taitse river, was deployed in the hills and had been partly moved to the northward. Part of the Fifth corps, which arrived here a short time before the fighting began, had been sent at once to the front, while the remainder had reported with the Second corps and was being held in reserve.

All the trees and Chinese graves on the firing line which in any way might obstruct the operations had been removed by the Russians.

On the morning of September 1 all the noncombatants were ordered to leave Liaoyang, and the Russian commissary of police sent an order to the Chinese magistrate of the city that all Chinese must leave Liaochang within two days. The Japanese were seen to be in possession of the Russian positions at Shushan, and it was learned also on the morning of September 1, that General Stakelberg had withdrawn during the night.

Chinese Panic-Stricken.

The Chinese at Liaoyang became alarmed and they began secretly to make Japanese flags. There was much confusion throughout the city. The proprietors of foreign shops made frantic efforts to sell their stocks on the streets at ridiculously low prices to anyone who would pay cash, or endeavored to have their property loaded on railroad cars. The flight of Japanese shells over the railroad station stampeded them, however, and they fled in disorder, glad to get away with their lives.

The foreign military attaches who were in Liaoyang were taken north under escort.

The war correspondents noticed, on this morning, an unbroken line of ammunition wagons, batteries of artillery and Russian troops using the railroad bridge to get out of Liaoyang to the north.

It became known that General Kuroki was making a dash around the Russian left for the railroad at a point fifteen miles north of here.

Takes Russian Defenses.

Kuroki crossed the Taitse river the night of August 30, and took the splendid Russian defenses at Fensihu with very little effort. He continued to advance steadily and turned the Russian left. His army covered fifteen miles in quick time. The effect of its presence and movements was to force the Russians to leave Liaoyang and retire in the direction of the coal mines east of Yentai.

Kuroki threatened to cut General Kuropatkin's army of seven corps in two.

The correspondents were eating lunch at the railroad station at noon when the first Japanese shells struck the foreign settlement. Within fifteen minutes' time all the troops, hospital attendants and merchants and other people in the vicinity had cleared out and only a few Chinese remained. The Japanese shells continued to fall in the settlement. Chinese, under the direction of Russian officers, were making hurried efforts to remove the stores and ammunition from the railroad sheds.

Shells Fall Everywhere.

The Japanese shells came from Shushan and a Russian shrapnel fire was soon turned on the battery there. One Japanese shell killed a horse as the animal was passing the telegraph office. Others burst in the postoffice, in the hospital in the Red Cross tent, which had been put up in the garden around the station and in the public park under the ancient pagoda.

The crowd of refreshment takers fled helter skelter with restaurant keepers and waiters, panic stricken officers and other orderlies, soldiers and the riff-raff of the streets, all seeking refuge beyond the northern wall of the city.

The Chinese immediately began to loot, but whenever they were caught immediate punishment was meted out to them. This was at noon of September 1. On that day the Russians filled all the trenches, rifle pits and forts to the west and south of Liaoyang with fresh troops. They kept up a continual shrapnel fire upon the Japanese approaches without being aware of the extra movements of the enemy. During this time the main body of the Russian army was in active retreat.

Could See Japs Advancing.

At 6 o'clock in the evening the shells were falling just outside the western walls and on the Russian heliograph station at the northwest corner of the city. From this station the Japanese could now be seen. They were advancing splendidly over the southern hills. Their formation was regular and open. The ranges of the Japanese position were known to the Russian gunners, who poured in a terrible and destructive fire upon the enemy. A Japanese battery and two companies of infantry were seen skirting the main south road. They were moving toward the city and finally disappeared behind Kowliang.

The positions at Chiafantun were seen to be deserted. The right flank of General Nodzu's army had previously taken up a position to the west of Chiafantun, whence his troops charged the Russian breastworks. One Japanese battalion lost every officer in this charge and a corporal was seen to take command. The Japanese intrenched in the vicinity of Tishi, near where General Ivanoff had had his headquarters the day before. The Russian Red Cross worked continually on the firing line.

The Japanese shells fell thick over the Russian battery located in a western suburb of Liaoyang.

Kuroki's Quick Move.

At 8 o'clock on the morning of September 2 the transport of the Russian army was still going out to the north and the head of the train was almost opposite Yentai.

The fighting around Liaoyang began at daybreak, when shells were thrown in from the entire Japanese line.

General Kuroki severed his connection with the Japanese center army and left the Anping road unprotected. His army was in front of the last Russian position in the coal mining district to the east of Yentai. The importance and effect of Kuroki's dash to the rear of the Russian army was plainly evident.

The foreign military attaches had been escorted from Liaoyang to the north. They left their train at a point near Yentai and proceeded in the direction of the fighting between Kuroki's troops and the Russians who were attempting to hold them back. Here the Russians lost an important hill position, only to gain again and lose it a second time.

Russians Are Discouraged.

This greatly discouraged the Russian soldiers. It is declared that the immediate precipitous retreat of the entire Russian army at Liaoyang was prevented by the retaking of this hill, which lies to the east of Yentai. The Russians, however, could not hold the hill and the Japanese were again soon in possession. General Kuroki's attack at Yentai was soon relaxed, however, and General Mistchenko and his command was withdrawn from the Russian defenders of Yentai and sent over to the west side of the railroad to make a demonstration and to aid in covering the retreat of the main army.

At 5 o'clock in the afternoon of September 2 two Russian regiments, which had been beaten to the east of Yentai retreated toward Yentai. They were ordered back to the fighting line while still under a sharp Japanese rifle fire. They met a large number of Russian troops who had been driven back from around Yentai. A train passed bearing 300 Russian wounded to the north. The road was filled with Chinese refugees from the coal mines as well as wounded men, all hurrying to the rear.

Night Ends Cannonading.

The cannonading came to an end when the sun went down. Throughout the night of September 2 it broke out again fitfully, but the periods were of short duration. The morning of September 3 showed that General Kuroki had not attained his object, but the victory of the Japanese over the Russians appeared to be complete. Russian troops and refugees from Liaoyang in rickshaws were pouring north along the railroad and along the main highway on their way to Mukden.

The Russian press censors, who were stationed at Liaoyang, Yentai and Mukden, fled to Harbin. Newspaper correspondents were instructed to send their telegrams to Harbin, which is about 340 miles up the railroad from Liaoyang.

The night of September 2 was marked throughout by furious rifle-firing outside of Liaoyang. The Russian troops were in a highly nervous and excited state. They were convinced that the Japanese were creeping forward irresistibly all the time.

Flames Visible for Miles.

September 3 saw the resumption of the cannonading at dawn, and the vast clouds of smoke which arose slowly from the burning warehouses in the foreign settlement at Liaoyang could be seen for a distance of ten miles. General Mistchenko, however, had been successful in clearing the enemy away from the Russian right flank.

General Zassalitch, of Yalu fame, with 1,000 men, on September 3 was smoothing the road north of Liaoyang for the retreat of the Russian artillery and troops. He was also trying to put into effect various devices calculated to delay the advance of the Japanese.

As a correspondent passed the fire-swept bed of the Taitse river on his way into Liaoyang at noon of September 3 the Russian infantry had retired from the immediate front to shelter behind the south wall of the city. Upon getting into Liaoyang it was seen that the Japanese rifle bullets were falling around the railroad station and that all the stored lumber and wooden buildings in that vicinity were in flames.

Japanese Enter Liaoyang.

Two hundred and twenty Chinese who had been wounded by the shells falling inside the city had been brought together at the Chinese Red Cross station and were being cared for. It was apparent that the Russians were using the walls of the city for protection. This the Japanese discovered, and consequently they continued to throw shells into Liaoyang. The number of Chinese wounded increased.

During the night of September 3 all the Russian sentries were withdrawn from Liaoyang. The pontoon bridges across the Taitse river were hurriedly removed and the Japanese entered and took possession of the city.

"Liaoyang at sunrise on September 4 presented a dismal spectacle. The Russian settlement was burning and overhung with a pall of smoke. The scared Chinese were in hiding and the worn-out Japanese were bivouacking in the suburbs. Not a shot was fired inside the barricade that day and not a Russian left except a few deserters disguised as Chinese.

"The town was not greatly damaged by the shell fire, but all the European shops and the wealthy Chinese residences had been sacked by the Tenth Siberian rifles.

When General Kuropatkin withdrew his forces north of the Taitse river in his retreat northward, General Stakelberg, with 25,000 men, remained on the south bank in disobedience of orders. For the time being he was completely cut off, but succeeded by desperate charges in regaining the main army.

Losses on Both Sides.

While the battle of Liaoyang proper lasted six days, the outpost engagements, leading up to the main event, covered a period of five days. In the eleven days of fighting the Russian losses were 40,000 men, while the Japanese lost in killed, wounded and missing, about 25,000.

Foreigners who observed the Russian troops at Liaoyang before the battle are of the opinion that the Russian reverse was the result of the previous four months of failure in the field; of what they termed the "demoralization," which began at the battle of the Yalu.

The task of General Kuropatkin seemed to be hopeless. The railroad during the month of July scarcely sufficed to bring into Liaoyang enough Russian troops to replace those who were lost by death, sickness or wounds during the month, and it was estimated that the Russians would have to outnumber the Japanese two to one in order to be successful against them.

Scenes witnessed in Liaoyang before the big battle gave evidence of a certain degree of demoralization on the part of the Russian forces.

Russians Forget Hardships.

After the Japanese had won their first successes at Anshanshan and Kaofengshi and were advancing rapidly on the second line of Russian defenses, some of the Russian officers who had been on the fighting line flung their responsibilities aside immediately after they reached Liaoyang and sought forgetfulness of the hardships of their past work and the dangers of the future.

The Pagoda garden was gay with merry-makers; there was much drinking and sounds of music and the singing of "chantant" songs in women's voices was heard from the officers' quarters in the foreign settlement.

At the beginning of the five days' fighting the hotels and public resorts of Liaoyang were deserted, soldiers and officers having been ordered to the fighting line. But as the men came back into the city from the front many of those who could indulged themselves in such pleasures as the city afforded.

Underestimated the Japs.

The Russian soldiers persistently underestimated the strength of the Japanese, and in Liaoyang the statements were often heard, even after the fighting at Haicheng and Anshanshan, that the war had not begun yet and the Japanese were playing a very dangerous game.

It was generally believed among the Russian troops that the Japanese were enlisting boys and old men in their ranks, for otherwise they could not account for the large numbers of soldiers opposed to them in the field. The attitude of many of the Russian troops in regard to the fighting ability and strength of the Japanese is changing, however, and the more intelligent are beginning to give the enemy full credit for his sagacity and worth.

Chinese Greet Japanese.

The action of the Chinese of Liaoyang, when the change in masters of the city occurred, was astonishing. They had fattened long on the Russian occupation, but as soon as the Japanese were known to be coming in they made ready with their newly manufactured Japanese flags to greet the conquerors.

Chinese shopkeepers at once hid all the stores in their possession which bore Russian colors. This was a wise precaution, as the Japanese soldiers later looted all such property.

Just before the Japanese entered the city many Chinese resorted to the fumes of opium, and as the Japanese soldiers marched through the streets of Liaoyang they splashed the mud and filth of the roadways upon Chinamen lying drugged and senseless in the low opium joints fronting on the narrow alleys and passages.

The Country Around Liaoyang.

To understand the conditions under which the battle of Liaoyang was fought, a knowledge of the topography of the country is desirable.

The general situation is determined by the Kingan ridge, which forms the backbone of Manchuria. This ridge runs from the south of Port Arthur right up to the great bend of the Amoor river, at Khabarovsk, and is, in fact, the cause of that bend, the Amoor flowing round it to the north. On the east of this backbone ridge lies the valley of the Yalu. On the west of the same ridge lies the valley of the Liao river.

From this main backbone ribs of low hills go westward into the Liao river plain. The railroad from Port Arthur crosses several of these ribs, and in parts it resembles a switchback railway. Between each pair of ribs a stream flows down into the plain, going generally due west and joining the Liao river in the valley. This valley is of immense fertility and is covered with magnificent fields, with numerous villages and farms and numberless tombs, each marked with a tuft of trees, generally elms, willows or firs and pines. It is recorded that the elms near Liaoyang are heavily decked with mistletoe.

The streams between the ribs are mountain torrents, liable to sudden floods from cloudbursts, and the bridges, some of M. de Witte's masterpieces, allow for these floods by the great height of their arches. Westward of the railroad, toward the Great Wall, are magnificently irrigated and cultivated plains, producing tobacco, cotton, millet, hemp and such fruits as pears, grapes, peaches and cherries.

The last rib of hills from the main backbone is some four or five miles south of Liaoyang, the backbone being in sight to the east of the town. The mountain stream corresponding to this rib is the Taitse river, which flows just north of the town in a wide, sandy bed, with numerous sand banks, one of which divides its stream into two main branches. The Russian position before their retreat began was a half circle, the southern front being on the rib of hills four or five miles south of the city, while the right and left wings touched the Taitse river, thus practically encircling the town of Liaoyang.

The Japanese position was south of the rib of hills and stretched in a larger semicircle outside the semicircle of the Russian troops.

The land between the southern rib of hills and the city is flat and richly cultivated. It is dotted with suburban villages, and is largely covered with market gardens and fruit orchards. To the east of the town the ground gradually rises till it passes through the foothills into the backbone ridge. Northward of the city comes, first, the wide sandy bed of the Taitse river, and then the flat, richly cultivated plain, twenty-five miles in extent, which divides the Liaoyang from Mukden. Westward of Liaoyang the country is generally flat, and extends in an open plain toward the main valley of the Liao river, into which the Taitse flows.

City of Liaoyang.

The city is built four square. It is surrounded by walls of stone, topped by brick and crenelated for archery or gun fire. The walls run north and south and east and west. There is a main gate in the center of each wall—thus, there is a north gate, a south gate, an east gate and a west gate. The north gate opens out on the Taitse river. The west gate opens on the railroad station and the Russian cavalry barracks, close to the station. From the south gate the old imperial road goes southward to Newchwang. From the east gate another imperial road goes eastward to Anping and thence to the Yalu and Korea.

The city measures two and a half miles north and south, by two miles east and west, or rather this is the extent inclosed by the walls. But the walled space, five square miles in area, is far too large for the hundred thousand Chinese and Manchu inhabitants, and much of it is laid out in market gardens.

The city itself resembles all Chinese walled cities, except that its streets have, for the last five or six years, been kept somewhat cleaner and lighted at night, by order of the Russians. There is only one slight elevation within the walls, on which stands the Imperial Treasury. The Buddhist temple to Kwanyin, the Goddess of Mercy, is the finest object architecturally, while the mission stations are the most interesting to westerners. These had some twelve hundred converts five years ago, but, as Liaoyang was the center of the Boxer movement in Manchuria, many were terrorized into relinquishing their new faith. The missionaries relate with pride how one of their number, Dr. Westwater, prevented the Russians from storming the city when they were putting down the Boxer uprising by armed force. Liaoyang used to be the capital of Southern Manchuria, and is still important, its chief industry being the distilling of a native liquor from the hemp and millet of the surrounding fields.

South of the rib of hills already described, and which bears the name of Shoushan, is another small plain, which in its turn is bounded on the south by another rib, of which Anshan, the "Saddle Hill," is the chief peak. Anshanshan, or "Saddle Hill Village," is noteworthy as being the most northerly point reached by the Japanese armies in the Chino-Japanese war of 1894-'95. At Anshan they made a feint at Liaoyang, which was one of the strongest centers of Chinese troops, and then turned southward to Newchwang.

CHAPTER XXXIII

THE DRAWN BATTLE OF SHA RIVER

Ten Days of Continuous Fighting with the Russians for the First Time on the Aggressive—Battle Line is Forty Miles Long—Kuropatkin Checks the Japanese Army after the Hardest Fought Battle of the War—Richard H. Little's Magnificent Description of the Spectacular Engagements—Both Armies Go into Winter Quarters, but Keep up Brisk Skirmishes.

GENERAL KUROPATKIN'S retreat from Liaoyang ended at the Sha river, near Mukden. The Russian commander intrenched himself there, and in a bloody ten days' battle succeeded in checking the Japanese advance. The battle began on October 6 and lasted until the evening of the 16th. It may properly be called a drawn battle, although the Russians had won a decided advantage by holding the Japanese in check.

Sha river was the last battle of the 1904 campaign, as both armies went into winter quarters there and kept up a skirmish campaign throughout the winter.

The most brilliant and detailed account of this great battle was written by Mr. Richard H. Little, the war correspondent of the Chicago News, with General Kuroki's first army. As it contains the best detailed description obtainable, I reproduce it herewith in full.

The remainder of this chapter is Mr. Little's eye-witness account of the battle of Sha river:

The ten days' battle along the Sha river was the hardest-fought battle of the Russo-Japanese war, before the winter closed in on the combatants. In every battle fought before this the Russians had determined beforehand to retreat or at least say they did; but in the ten days' battle the Russians were attacking and doing their utmost to go forward, while the Japanese seemed to think the attack of the Russians an opportunity to push on and capture Mukden.

Neither side was victorious. The armies simply clinched and hung on until exhausted. When they fell apart it was found that each side was about where it started, except for the terrible slaughter. I do not know the Japanese losses, but I know that the Russians lost enough men to take the heart clear out of any army but the Russian. The censor at Mukden permitted me to cable that the Russians had lost about 40,000 men, 32,000 missing and wounded and 8,000 dead. Since then I have been told by a Russian medical officer in a position to know that 35,000 wounded passed through Mukden, to say nothing of those not yet brought in, and it may be that 50,000 is nearer the correct number of the Russian casualties than 40,000.

The Japanese are not as frank about their losses as the Russians. Since the battle was so nearly a tie and since the Japanese attacked almost as much as the Russians did, it seems to me that the Japanese must have lost not far from as many men as the Russians. As for the loss of guns, about which there is so much contention, I believe from as thorough an investigation as I have been able to make that the Russians lost more than thirty guns and the Japanese sixteen. The end of the battle left the belligerents still defiantly facing each other along the Sha river and from any of a dozen hills the unique sight is presented of both armies in plain view, no longer fighting but digging intrenchments for dear life.

Both Claim a Victory.

The ten days' battle settled the question of the winter campaign. The Russians will settle down in Mukden for the winter while the Japanese go into winter quarters at Liaoyang. Both armies are now out in the open, where it will be impossible to remain through the bitter cold of a Manchurian winter, and it will take another battle to decide where winter quarters are to be taken up. The Japanese claim the ten days' battle as a victory, because thė Russian advance was checked; but the Russians in turn checked the Japanese advance, and besides the Russians point out that their advance was only checked—they were not defeated or turned back—and that they secured positions that will be of the greatest value when the advance is renewed.

So the Russians will, in all probability, again move forward, or if they do not then the burden of attack will be thrown on the Japanese, so that it will take another terrible battle to decide who had the advantage in the ten days' affair.

Saw Battle from Mukden.

If a modern battle is hard to see the battle of the Sha river was hardest of all. Here was a line of battle more than forty miles long. One end of it was out on flat plains along the railroad, the center was in the hills, while the left wing fought furiously against nature as well as the Japanese in the mountains. The battle raged every inch of the way from ten miles west of the railroad to Pensiho, some thirty miles west of Yentai, and well up in the mountains. The part of the battlefield in the plains was only twelve miles south of Mukden, and every day of the fighting the great walls of Mukden were crowded with a curiously mixed assortment of Russian officers and soldiers and Chinese mandarins and people of all classes, who watched with mingled feelings the bursting of the Japanese shells and listened to the reverberations of the artillery, so heavy that the window glasses in Mukden rattled as if shaken by an earthquake.

The weather during the first few days of the fighting was the same kind of glorious fall weather that a kindly providence sometimes vouchsafes to us in America, but up in the mountains the nights were very cold and I remember when we started with Rennenkampff for Pensiho our horses' feet broke the thin ice coating of the little creeks. The terrible bombardment of the first few days brought the rain down in torrents. During the remainder of the battle a cold, merciless rain that turned the roads into bottomless quagmires increased tenfold the miseries of the soldiers, weakened by the long days of ceaseless fighting and the weary vigils of the longer nights.

Battle Not Spectacular.

It was not a spectacular battle, any more than a bulldog fight is spectacular. The armies simply locked and held on; yet up in the hills and mountains one could get vantage ground and see sights that were worth a lifetime. Down on the plains the shells were here, there and everywhere, before us, behind, this side, that side. Finally, in disgust, after having been shelled out of three places selected with great care and discretion, our small party turned its back on the plains and retired to the friendly hills in the east.

I was in at the beginning of this great struggle and for five days watched Rennenkampff straining every nerve to turn the Japanese right flank and capture Pensiho. All of which is a story all by itself. I was in at the death, too, and saw the 10th corps hold fast in a hell of shrapnel and shimose, hold fast and then afterward in the night and the cold and the pouring rain crowd forward and push the Japanese back. This is likewise a separate tale and as such will be duly set in its own place.

Kuropatkin Nerves Soldiers.

We knew in Mukden that the Russians were to advance, although the secret was most jealously guarded. A general told me that Gen. Kuropatkin would much rather have had the Japanese advance so that the Russians could have fallen back to Tie pass in accordance with the general's plan and then made the start forward from that point; but inasmuch as the Japanese had stopped short at Yentai and gave signs of going into winter quarters at Liaoyang the Russians were forced to attack at once. General Kuropatkin's proclamation was read in all the various corps and detachments of the army, October 5. It briefly reviewed the campaign, spoke of the treachery of the Japanese in attacking Port Arthur before the declaration of war, declared that the so-called Japanese victories were gained only because the Russians had beforehand decided to retire at each point after checking the enemy's advance, and then called on the soldiers to march bravely forward to the relief of their heroic brothers struggling in Port Arthur.

Everywhere solemn mass was said by the priests in the presence of the troops paraded in hollow squáres and after the commanding officers had addressed the troops and read the proclamation the men threw their hats in the air and cheered, the bands played the national anthem and there the greatest enthusiasm prevailed.

It was a cool October day. In the fields the kiaoling (millet) was stacked just as corn is stacked in the United States, and looking just the same. The gray mist hung over the mountains, which crowded in close on us all around, and the sun, which had been under the clouds all day, burst forth just as the priests began the service and shone down brightly. The priests and a choir of soldiers chanted the service.

Troops March Cheerfully.

General Rennenkampff, muffled in his big woolly "bourke," the black cape that covers its wearers from the shoulders to the ground, stood on a little eminence, and as the troops filed past inquired in the fashion of the Russian army, "How are you, my children?"

The men shouted back, "Very well, your excellency, thank you," with a vim that made it sound like a college yell. It was cold, but everybody was happy. We were going to Port Arthur.

All that forenoon we marched steadily. The sun came up and the mist disappeared and finally we grew warm again. We marched in a southwest direction down the valley for several miles, and then began climbing over the mountains. The roads were hardly more than trails and sometimes very steep and rocky, but the troops were fresh and besides it's a long way to Port Arthur and we had to hurry. At noon we halted at a Chinese town and remained there until the next morning, when we pushed on to Santranitze, a little town only a few miles from Pensiho.

Here we had to wait until we heard from the 3d corps on our right. The 1st, 2d and 3d corps were moving down from their former positions on the high road running straight east from Mukden to Feushan. All the various corps and divisions of Kuropatkin's army were to move southward in concert, but Rennenkampff's cavalry had moved a little faster than was planned or else the other corps were slower than it was thought they would be. At any rate we had to wait at Santrantze while Japanese scouts peeked at us from the tops of the mountains and smiling Chinese wandered out to inform the enemy how strong our force was and of our sinister designs in regard to Pensiho.

Troops Under Fire.

The skirmishers of the main column across the river had now engaged the enemy and taken all the attention of the Japanese and we recrossed the Taitse in peace and immediately afterward put the horses to a gallop and rode under fire to the protecting walls of a Buddhist temple which stood a short distance outside the town. The men dismounted, every third man remaining behind to hold horses, and the men went immediately into action.

The advance was made by rushes until we reached the town. The Japanese seemed to be across a small river at the foot of the mountains at the other end of the village. The Cossacks took shelter wherever they could find it and opened a hot fire. The Japanese began volley firing and seemed inclined to stick to their position. A big Cossack bounded to his feet with the blood gushing from his breast. He made the sign of the cross on his forehead and fell dead. Another Cossack dropped his gun and grabbed his wrist, where a round red hole suddenly appeared. A man suddenly sat up in a dazed way and commenced feeling cautiously about his head as though in doubt as to whether that deep red furrow over his ear was really a wound or whether he was waking from a bad dream.

Frightful Losses Around Pensiho.

On October 8, at daylight, Gen. Rennenkampff advanced on Pensiho and at noon he reached the little town of Ounion, at the base of a number of parallel high mountain ridges that separated it from Pensiho. In two hours he had driven the Japanese skirmishers from the woods beyond the village and up the mountain and occupied the first range of hills. For five days the fight raged around Pensiho. Rennenkampff's men did all that men could do. The Russian losses in the rugged cliffs around Pensiho were frightful. The regiments were fairly hurled against the Japanese and only the battered remnants came back. The little streams that came tumbling down the mountain ran red with blood. But it was in vain.

On the night of October 12, while the Japanese shells were falling freely among the transports and the marching columns Rennenkampff turned his back on Pensiho and marched northward.

Both Attack by Night.

The Japanese seemed to have arrived at this conclusion some time ago, for their infantry attacked repeatedly at night at Liaoyang. The Russians have discovered the truth of this assertion also, for during the last days of the Sha river battle they hung tight to their positions in daytime and at night hurled their infantry at the Japanese. It was a night attack that gave the Russians the sixteen guns they captured. Two or three miles east of the railroad is a conical-shaped hill, with gently sloping sides. At the top of the hill in the exact center is a little tree. On this hill the Japanese artillery was planted. It did deadly work and silenced and drove back all the Russian guns in range.

Several attempts were made to take it in daytime, but the only result was frightful loss in the lines that vainly tried to weather the awful storm of shrapnel. Finally, after dark on Sunday night, the 9th and 20th European regiments, with a mixed brigade from the 3d and 4th corps supporting them, advanced against the batteries on the hill. The guns worked with lightninglike rapidity, but in the darkness the Japanese gunners lost the deadly accuracy of their aim.

Storming of Lone Tree Hill.

The two regiments pressed on, although they were losing heavily. The Japanese gunners stuck to their guns even when the Russians came over the intrenchments. The infantry support came to the rescue and a hand-to-hand fight ensued. There was no more firing. The much-vaunted guns and the repeating rifles and all the triumphs of modern inventive genius were silent. It was the time for the cold steel, for the archaic sword and lance and bayonet. It was stab and hack and parry and thrust. In the darkness friends and foes were mixed in one yelling frenzied mob. The Japanese wore white bandages on their arms to distinguish each other from the enemy, but it also told the Russians where to strike. Each Russian struck and fought at every white bandage he could see. A Russian sergeant stabbed in the arm and falling beneath the guns bound a handkerchief about his wound and plunged into the fight again and his own captain cut him down and would have run him through but he cried out in time and tore off the handkerchief that had marked him as a Japanese. The fury of the Russian attack was too much for the Japanese. They fought like brave men, but they were fighting men just as brave and much stronger. Those who could crawled out of the melee and fled down the hill.

The Russian guns, which had followed fast on the heels of the infantry, took position and the hill was won; but the cost—aye, the cost. It was a big price, but the Japanese paid heavily, too. At daylight the sum could be added up. There they lay in the trenches and scattered all about the hill—Russians and Japanese, some of them still in a death grip. A Cossack had both hands clinched in the coat of a Japanese, whom he had been trying to drag down. The Cossack's head and body were covered with the deep gashes the Japanese had given him with a sword, but the man had held fast until a comrade had run his bayonet through the Japanese. Nor even then had he relaxed his hold, nor would he in death, and he was buried with the coat of the dead Japanese in his hands, for the burial party could not pry loose his grip. Three all but headless bodies lay near the guns, which were daubed and splattered with blood. Nine hundred and sixty-two Japanese and more than 500 Russians were buried on the hill where the guns were captured.

Regiments Stood Firm.

The weakest place in the Russian line was left of the center where the 4th corps had been spread out with too wide a front. The Japanese found the weak spot speedily and pushed hard against it. At the same time the Japanese pressed a vigorous attack against the 10th corps along the railroad. A part of the 17th was detached to help Stakelberg and portions of corps were thrown in to re-enforce the 4th, which, although pushed back some distance, succeeded in holding its place and preventing the Japanese from breaking through. The 10th and 17th on the right were being terribly pounded.

Again and again they sent to Kuropatkin for help, but he had no troops to give them. He only sent back word "You must hold your positions at all hazards." They held. The 17th was terribly punished, whole regiments were cut down until they were hardly decent-sized batteries and the corps lost twenty guns. On Monday all along the line, but especially upon these two corps, the Japanese made a tremendous artillery assault and followed it up in the evening with an infantry advance that was turned back.

That night in the sticky mud, with the rain beating in their faces, faint from hunger and exhaustion, the 10th corps moved steadily forward. The Japanese were ready and poured in a very hot fire. The 10th stumbled and staggered through the mud; it moved very slowly, but the flashes of the rifles showed that it did move. And always forward. Not far, to be sure, but ahead of its old position. It gained ground. It stopped there and held the place until the end. There's good stuff in that 10th corps.

Japanese Fire Accurate.

The terrible loss on the Russian side was due not only to the fact that during most of the battle they were the attacking party and suffered enormous losses as they came out into the open from the Japanese infantry and artillery, well placed in positions almost or entirely hidden, but most of all to the terrible fire of shrapnel and shimose shells that the Japs poured on all parts of the Russian line day and night. The Japanese did not merely content themselves with shelling the Russian firing line; the infantry and the mountain and magazine guns looked to that while the shrapnel roamed all over the field, now hammering down on the troops marching forward to re-enforce the fighting line, now pouncing down on men huddled behind a stone wall in reserve, now bursting exactly in the middle of the troops lying in support. The rear of the Russian army lost almost as much as the front. The great shimose shells seemed to concern themselves particularly with the rear.

When we were standing in the compound of the Chinese house, where the headquarters of the 10th corps had been established and which was supposed to be in the rear, shrapnel continually dropped in the road not fifty yards away. We would have gone back, but shimose shells were falling across our way of retreat, some close at hand and others so far back as half or three-quarters of a mile.

Three different times that day, in company with a Spanish and a German military attache, did we take up what we considered safe positions from which to see the battle, only to find that shimose shells were passing over our heads and that we were far inside the danger zone. In despair we retired in good order to a village a mile away from the headquarters of the 10th corps that we might have tiffin in peace. We had not finished boiling water for the tea before a shell fell across the road and burst in a group of cavalry that had supposed it was as safe as though it had been back in Mukden. There was a dull boom back of our house and looking we saw the pillar of mud and rocks and smoke that a shimose shell hurls into the air.

Japanese Broaden "Danger Zone."

Coming back that morning from the headquarters of the 10th corps I was trying to get a picture of a Japanese prisoner when shimose shells began to hurl up mud geysers all around. A shell struck fair in the center of a transport train and horses, mules and men seemed to go into the air. Four soldiers who, perhaps, were congratulating themselves on getting out of the fight and were boiling tea over in a kiaoliang (millet) field were instantly killed by a shell that seemed to fall directly in the center of their camp. Shells were falling behind all the walls and buildings in the town behind us. Through a glass one could see the soldiers of the reserves and supports, obliged to stay where they had been stationed, turning confusedly this way and that, lying down, standing up, then running a few steps forward and back again and acting like wild animals surrounded by a prairie fire, blinded, dazed, not knowing which way to turn and waiting helplessly for death.

Military experts speak of the "danger zone" of a battlefield. The Japanese have broader ideas of the danger zone. They have extended it to include every acre of ground occupied by their enemy, firing-line supports, reserves, extreme front, middle distance, extreme rear. Watching their shells, it was easy to see that the Japanese had surveyed all the country in front of them and had the range of all buildings, roads and ravines where troops not in the firing line would find shelter.

They had apparently made calculations that in one building headquarters would be established, that a long stone wall with a ravine behind it would be where reserves or supports would find shelter, that on this hill near a little clump of trees batteries would without doubt go into action. They had the exact location and range of every place. We had abandoned the road entirely. For a mile, although the road was not a straight one and took several bends and curves, it was dotted with exploding shimose shells.

Japanese Plans Come Out Right.

The artillery wagons and the carts taking supplies of ammunition and food to the front inquired where the road was only that they might carefully avoid it and go forward instead by bumping straight across the fields. It was usually unnecessary to ask about the road. Fountains of mud marked it much plainer than sign boards would have done. By noon new roads had been worn off to right and left of the main road, but by 2 o'clock, thanks probably to Chinese spies, the Japanese artillery had found these, too. It was a terrible pounding that the Russians had to take. The Japanese seemed to have anticipated the Russian advance and they prepared for it by figuring out what positions the Russians would take and putting batteries in place to shell each spot. The Russians sent back shrapnel for shrapnel, but they had nothing to offset the deadly shimose fire.

Russians Shell Own Battery.

Such a sweeping general rain of Japanese shells, besides killing and maiming such an immense number of men, also caused considerable confusion in the Russian lines. One Russian battery by mistake began shelling another battery. The second battery should have known that Japanese shells could not have come to them from such a direction, but the mere fact that shells seemed to fall anywhere from anywhere had prepared the Russians to be surprised at nothing, so the second battery immediately took up the challenge of the first battery and the two engaged in a deadly duel that only ended with the destruction of the second battery.

It seems impossible that any army could suffer the loss the Russian army did along the Sha river and still hold its position, keep up the desperate fight as long as the other side did and at the last actually move forward in places and occupy advanced territory. I watched the terrific shelling of the 10th corps on Monday and could not understand why the troops did not suddenly break into headlong retreat or at least retire. Such a red field of slaughter has seldom been seen on any battlefield.

The Russians took their punishment without the slightest sign of giving way. The shells pounded, pounded, pounded, left, right, center, now a whole line of shimose fountains straight across the fields, now a dozen puffs of white smoke, as the shrapnel leaped down on this place; then swung back and staggered the column, marching over in a field; then leaped over, as if to devour a wagon train trying to slip along unobserved in the underbrush along the creek. The German military attache bit his lip and kept opening and shutting his field glass and murmuring in German. The Spanish attache was continually clapping his hands and crying, "Bravo, bravo," as though he were watching a play. I shouted, "Hold them, 10th!" It was like a football game, where our team is on its two-yard line, with the score tied and three minutes left to play.

The 10th held. It lost men by the hundred. Corpses and bleeding men and men rolling on the ground shrieking their death agony were everywhere, but the 10th held. We sat grim and tight in all that hell and when the Japanese tried to follow up shrapnel and shimose with the bayonet drove them back. Then in the blackness of the night, in a cold rain, that fell in torrents and in the face of a pitiless rifle fire, the old 10th, or rather what was left of it, rose up and went forward to a position a full half mile or more in advance. They held there, too. It is a great thing in an army that it never knows when it's beaten.

CHAPTER XXXIV

THE SIEGE OF PORT ARTHUR

(Continued from Page 318.)

General Nogi's Land Forces Continue the Attack from the Rear and Take Fort after Fort, but with Heavy Loss—Dynamite Bombs and Bayonets Used in Fierce Hand-to-Hand Conflicts—Two Months of Fighting for 203-Meter Hill, the Key to the Russian Situation—Stoessel's Defense of the "Russian Gibraltar" the Most Gallant in War History.

DURING the evening of August 20 the Japanese left flank cut the entanglements on the slope east of Keekwan Fort hill, and before dawn, August 21, a center division of the Japanese gained the foot of East Panlung fort, called by the Japanese East Banjusan fort, and tried to cut the entanglements, but the troops were repulsed. At daylight all the Japanese batteries of the center division concentrated their fire on the Banjusan and Keekwan forts, but despite the artillery support the Japanese were driven out of the East Keekwan fort by an attack of a massed body of Russians from the Chinese wall.

The assault threatened to result in an utter failure to secure a foothold on the fortified ridge, and General Nogi summoned the generals of the left and center divisions to consult with him concerning the operations.

During his absence at 11 in the morning, without special orders, several companies of a regiment of the center division emerged from their trenches at the foot of the East Banjusan fort, and in tens and twenties charged up the slope to the broken wall of the trench around the crest of the fortified hill, under cover of a splendid shrapnel practice from their field batteries in the valley.

The Russians behind the wall and in the fort poured out a hail of rifle bullets and the machine and quick-firing guns belched forth on the intrepid Japanese. Twice the latter were forced back till the slope was covered with dead bodies, but a third attempt was made and a score of Japanese reached the broken wall. The terrible fire of the Russians, however, started their retirement.

Heroic Sacrifice Brings Capture.

Suddenly a Japanese officer, regardless of danger, stood up, called out an order, planted his regimental flag on the wall and was immediately riddled with Russian bullets. The effect on the Japanese of this sacrifice was instantaneous. The retreating infantrymen stopped, hesitated and then charged back, fought like demons, jumped over the wall, charged the Russians with bayonets and forced many of them up the glacis. over the ramparts and into the fort.

The Japanese were reinforced from time to time by rushes made by their comrades up the deathway from the trenches, and they stubbornly held the corner wall and a small section of the Russian trenches till 5 o'clock, when two companies of another regiment worked through the trenches and attacked the West Banjusan fort, the magazine of which had been blown up by a shell.

Capture West Banjusan Fort.

Taking advantage of a diversion of the Russian fire, the Japanese attacking the first fort forced the Russians back, captured the fort at the point of the bayonet and pursued them along the connecting trenches to the Chinese wall. The West Banjusan fort was captured at 7 o'clock after a slight resistance, the Japanese being unable to occupy it, as the magazine was blown up, but they held the trench line around the crest.

In the afternoon three battalions of Japanese reserves were added to the center division and during the night a regiment left the division and assaulted and captured part of the south Keekwan fort.

At daylight August 23 the Russians concentrated shrapnel fire on the captured forts and the Japanese south of Keekwan were ejected from their positions. The Banjusan forts were subjected to an awful fire, but the Japanese in possession of them continued making bomb-proof trenches on the hills.

Slaughter Japs by Searchlight.

Later the Russians, anticipating an assault, made a strong counter attack at night. The Japanese advance lines were driven from the forts and were furiously attacked by large bodies of Russians, who also worked down on either side of the captured Banjusan forts to the valley, in an attempt to cut off the retreat of the Japanese from two forts. The Russian move was cleverly planned and their artillery commanders were well informed, for they opened fire with shrapnel on the advance of the Japanese supports across the valley, using starlights and searchlights with good effect. The general assault then began and for twenty minutes the whole Japanese line was engaged.

The left division from the trenches in front of the Keekwan forts charged the fortified hills. Part of the advance had gained the crests and shouted "Banzai" when their triumph was cut short. Suddenly the two powerful searchlights of the east forts lighted up the Japanese lines and Russian rifles and machine guns poured a deadly hail into the ranks of the attacking troops.

Destroy Lights and Silence Guns.

Despite the awful process of annihilation the Japanese stubbornly held the positions gained and their machine guns quickly located and quieted the Russian quickfirers. Finally the Japanese were slowly forced down the slopes to the trenches. On the west flank the searchlights of Etse and Tainyangkow forts played along the Japanese trenches, preventing the movement ot troops. The Japanese artillerymen concentrated their fire on the searchlights. Suddenly the light of Etse fort disappeared and the light of Tainyangkow fort followed.

The Japanese infantry on the right flank advanced on the trenches close to the Russian lines when the Etse light was flashed out in the faces of the advancing troops and Russian machine guns swept their fire along the lines, while rifles blazed with continual rattle. As the Japanese machine guns came into action they were located by the Russian starlights and made more distinctly visible by the searchlights, but they silenced the Russian quickfirers.

Meantime the fight was fiercest in the center of the Shuishi valley. The Japanese captured the Banjusan forts, outflanking the Russian forces and slowly forcing their lines back to the foot of the fortified hills. The Japanese were well supported, rushed forward furiously and engaged the Russians who had gained the valley in an attempt to outflank the forts. Though the Japanese plan of attack was destroyed by the Russian counter attack the Japanese fought with splendid determination. Slowly the Russians were driven back up the slopes, fighting desperately with rifles.

The Japanese artillery was unable to fire at this point as the Japanese forces were mixed.

Russians in Dark, Japs in Glare.

The skillful working of the Russian starlights and searchlights were utterly unexpected and bewildering. They never failed to locate the Japanese lines, which offered splendid marks for the Russian rifles and machine guns and rendered the Japanese machine guns of little use, as they were located and silenced by quickfirers before they could do any execution.

The Russians along the whole line fought in the blackest darkness and the Japanese with the most dazzling light in their faces. The rattle of musketry, the thundering of the Russian guns, the purring of the machine guns, the bursting of starlights and the flashing of searchlights along the whole line were wonderfully impressive.

The fight lasted continuously for six hours, till dawn, when the outflankers were forced back over the captured fortified hills followed by the Japanese, who captured the forts and joined their forces from the valley.

Capture New Banjusan Forts.

Before day August 22 the center division made a third attempt to capture the higher fortified ridge of Ash hill, the new Banjusan forts on the higher ridge, and the rear forts. The Banjusan forts were stormed and captured, the Japanese pursuing the Russians. Two hundred Japanese officers and men were killed. All belonged to a regiment of the center division which reached a neck between the double peaks of Wantai.

The Japanese captured a fort where they intrenched themselves and remained under an awful concentration of shrapnel fire. With daylight the Japanese artillery started a terrific bombardment. An avalanche of shells swept down on the eastern fortified ridges, quieting the Russian guns. The storm of bursting shells lasted half an hour. Over 400 guns roared in rapid succession until the entire ridge was enveloped in smoke.

General Assault Proves Failure.

For a time every Russian fort was silenced. When the smoke cleared away it was seen that the left division had advanced through the captured forts. One regiment attacked the east Keekwan fort, but was repulsed. Another regiment advanced up the northeast slopes of Wantai hill in an effort to effect a junction with the regiment which held the neck between the two hills.

The capture of Wantai hill meant the dividing of the eastern fortified ridge. A center supporting regiment was unable to scale the heights and was forced to intrench at the foot. During the night of August 24 the ranks of the Japanese were so decimated by the furious shrapnel fire of the Russians that they were forced to retire to the valley below the captured forts and what might have been a successful general assault, with the capture of the fortified ridge east of Port Arthur, was converted by the Russian tactics into a repulse, redeemed in part by the wonderful fighting qualities of the Japanese infantry.

Japs Lose 14,000 Men.

The Japanese casualties from August 19 to August 24 were 14,000. The center division alone lost 6,000 and a single regiment lost 2,500. Only six officers and 200 men of this regiment were left after the fight.

The retention of the Banjusan forts gave the Japanese a foothold on the fortified ridge as a result of six days of general assault. Heavy casualties marked this hazardous attempt to take a strong modern fortress by assault after a few days' investment. The ability of General Stoessel and the determined spirit of the Russians had been underestimated and the experiment, though successful, was never repeated. The Japanese army settled down to sapping and making parallels.

Russians Finally Give Up Forts.

For two weeks the Russians were unceasing in their efforts to recapture the Banjusan forts. They bombarded and assaulted by day the bombproofs and trenches. Though the Japanese resorted to decoy trenches and bombproofs the daily loss was 100 in each fort until September 8, when the Russians ceased their continuous effort to recapture the forts.

The captured Banjusan forts enabled the Japanese to mass a force under cover on the fortified ridge, and so using the Urh forts as a pivot, swing the line, capture the fortified ridge and envelop Port Arthur from the east. The capture of Shuishi was necessary to enable the construction of parallels to the Urh forts. The strength of the western fortified ridge was due to the fact that the Sungchow, Taiyangkow and Chair hill fort groups, with the advance fort on 203 Meter hill, formed the right angle of a triangle, the base line being from Etse to 203 Meter hill.

Refuse Quarter in Sorties.

During the operations from August 25 to September 18 the Russians made sorties and attacked working parties almost every night, while guns bombarded by day. As the Japanese trench line neared the entanglements on the fort hills the sorties became more frequent and determined. Quarter was neither asked nor given, the fiercest antagonism was displayed, and even stretcher-bearers were killed. The cause of this was that the Russians alleged that their forts were bombarded while a messenger under a white flag was delivering the emperor's message August 16. The feeling was augmented and later all flags were unrecognized.

The pioneers suffered heavy casualties in cutting entanglements. After the failure of the regulation devices for removing wires the pioneers were sent to attach ropes to posts to which entanglements were attached and troops in the trenches dragged posts and wires away. This worked satisfactorily until the Russians began the use of bracing wires. Then pioneers advanced and fell as though killed close to the entanglements and remained motionless until, unobserved, they could work along on their backs under the wires, which they cut with long shears. As a result of this expedient the Russians made certain that all were dead who fell near the entanglements.

Dynamite bombs were used by the Russians against the Japanese trenches and advance works. For assaulting forts, where it was impossible to throw bombs, wooden mortars were made, bound with bamboo. These were carried by the soldiers and with them bombs were thrown fifty to one hundred yards. These were the most effective of the many devices tried by the Japanese.

In order to recover the wounded when possible volunteers crawled from the trenches at night and worked along on their stomachs, pulling the wounded slowly to cover by the heels. Many of the wounded were not recovered.

Make Second General Assault.

While the siege operations following were in progress re-enforcements were added to the existing units and it was announced from headquarters that the siege operations would be finished with a general assault September 19 on the Namaokayama ridge, the 203 Meter hill, the half moon forts of the Shuishi valley, and the redoubt at the foot of Rihlung mountain. The bombardment was started at dawn. The Russians failed to reply till the afternoon, which delayed the attacks until 5 in the evening. Three battalions of the center division advanced from the parallels and attacked a redoubt at the foot of Rihlung mountain, which the Japanese twice before had tried to capture.

The Japanese were met by a tremendous fusillade from rifles and machine guns and were forced to retire to their trenches, but at 4 in the morning the main attack was made and the Japanese captured the redoubt, entering it through a breach made by the artillery. A fierce hand to hand encounter inside the redoubt followed, during which dynamite and hand grenades were used to destroy the blockhouses, whose brave defenders stood their ground until the last man was bayoneted. The Russian trenches were captured and occupied by the Japanese, but the redoubt was destroyed with explosives. The defending force consisted of four companies. The Japanese casualties were over 1,000. The Russians left only a few dead.

Capture Half Moon Forts.

The evening of September 19 a regiment division attacked the Half Moon forts in the Shuishi valley from parallels fifty yards distant. The first assault was made before dark. A battalion and a half advanced against the strongest west Half Moon fort and two companies moved against the east fort. Both attacks were repulsed. Twice again during the night the Japanese made unsuccessful assaults on these forts.

Early the morning of September 20 the whole force was concentrated in a furious assault on the west Half Moon fort. The moat was crossed with scaling ladders, and dynamite and hand grenades were used to destroy the Russian trenches and bombproofs. The fighting, which lasted twenty minutes, was carried on at the point of the bayonet. Finally the Russians were driven through the connecting trenches and the other Half Moon forts were all captured after slight resistance.

The Japanese casualties were 400. The Russians left but few dead. All the Russians' guns were captured, but they had been destroyed.

Slaughter Japs on Meter Hill.

The attack on 203 Meter hill and the ridge immediately north of it, called Nahaokayama, was started at 5 o'clock the evening of September 19 by two regiments. The first regiment advanced against Nahaokayama from trenches close to the Russian lines. In spite of the furious bombardment with which the Japanese supported the assault the Japanese infantry could not advance beyond the "dead ground" and were compelled to spend the night on it, close to the first Russian trench line. Meantime a second Japanese regiment advanced against the west slopes of 203 Meter hill and a third regiment moved against the southwest slopes. The second regiment was unable to make much progress, but the third regiment reached the foot of the slopes. During the night a party of sappers cleared away the Russian wire entanglements.

At 11 o'clock September 20 a small party of the second regiment charged forward to gain the foot of the west slopes. The Russian batteries were evidently waiting for the Japanese, as a most wonderful shrapnel fire was opened on them. Every man seemed to fall. Within ten minutes all the Japanese were down, either wounded or killed.

Seize Ridge by a Rush.

The artillery duel continued until 5:30 in the evening, when the first Japanese regiment, carrying its flags, advanced up the slopes of Nahaokayama to the dead ground, which the single company had gained the night previous. The Russians had retired to their first line of trenches and several companies of Japanese were extended along the slopes under cover of the brow of the ridge. With a rush they carried the trench line on the crest of the hill. The last rush of the Japanese was a splendid spectacle, ending with a bayonet encounter with the Russians on the full sky line. Stones, bayonets, swords and hand grenades were used by both sides. The utmost ferocity was displayed. When the Japanese had carried the east half of the ridge the Russians retired to the west half. Before darkness the Japanese were in possession of the whole ridge, which they retained despite an awful shrapnel fire from the Chair hill, Taiyangkow, Tiger's Tail and Liaotie forts.

Fail to Take Nahaokayama Hill.

At 10 in the evening the third regiment advanced on the southwest slopes of the advance fort on Nahaokayama hill. A small party gained the fort on the crest, but was driven out the next day. Four assaults were made by the Japanese during the next two days, but all were repulsed. The third regiment was finally compelled to retire from the trenches at the foot of the hill by a concentrated shrapnel fire from the valley below.

The unsuccessful attempt to capture the advance fort resulted in 2,000 Japanese casualties. A hundred and eighty Russian dead were found at Nahaokayama, which was covered by a network of trenches and bombproofs.

Japs Forced to New Tactics.

The capture of the redoubt enabled the Japanese to work parallels eastward, and possession of the Shuishi lunettes enabled them to work westward. With two forts on the west side captured and the two Banjusan and the three Keekwan forts on the east side, the attackers possessed a foothold on the front extending over more than half the eastern side of the fortified ridges. Nahaokayama ridge was a splendid basis for work by the parallelers on the 203 Meter hill and Chair hill group of forts, on the west fortified ridge. Every inch the Japanese now gained was by laborious sapping against the determined opposition of the Russians, who fought with tremendous earnestness, making sorties every night against the sappers.

The morning of October 10 there was a heavy bombardment of the east Urh fort, under cover of which infantry advanced and captured the trenches at the foot of the slope.

The night of October 11 three companies of the right division captured a smaller railway bridge on the northwest slope of the west Urh fort and also tried to capture a larger bridge at the foot of the west slopes, but they were repulsed. The night of October 12 a company of the center division made a surprise attack on the same bridge. The Russians retired in confusion, leaving their kits and overcoats. Later the Russians made a sortie and tried to recapture the bridge, but were unsuccessful. Simultaneously there was a sortie on the east Keekwan trenches, but the Russians retired after an hour's fighting.

Charge and Take Trenches.

The fire of the eleven-inch howitzers was so effective against the war vessels that the Russians resorted to placing a hospital-ship in the direct line of fire. The West Urh fort was damaged October 13. The battleship Peresviet was struck by several shells from the howitzers and caught fire. Golden hill fort was greatly damaged.

October 16 the Japanese attacked an intrenched hill called Hachimake Yama, between the east Urh and west Banjusan forts. Under cover of a tremendous bombardment the companies of the center division emerged from parallels at the foot of the hill and charged the glacis with bayonets and captured the trenches at the crest which the Russians evacuated.

It was announced October 25 that the following day there would be a bombardment of the east fortified ridge, from the West Urh fort to the South Keekwan fort, and also the Chair hill forts, to be followed by infantry attacks for the capture of the trench lines and glacis of the East and West Urh forts, the East and Southeast Keekwan forts, with demonstrations on either flank.

The bombardment commenced early in the morning, increasing in volume until 5 in the evening, when two regiments charged. When the first regiment gained the trenches at the crest of the glacis of East Urh fort the Russians exploded a mine. Pieces of timber, earth, stones and some Japanese were hurled hundreds of feet. The trench lines on both hills were captured by 6 o'clock in the evening with 250 casualties.

The rapidity with which the Japanese excavated trenches and made connections with parallels under fire was marvelous. They seemed fairly to sink underground.

Between 3 and 5 o'clock the morning of October 27 the Russians made determined sorties on the Japanese trenches near the West Urh forts and Hachihake Yama. The Japanese casualties were 300. Six officers were killed and eight wounded. The Russians left sixty dead.

The same night between 9 and 2 o'clock the Russians made four sorties against the captured trenches near the West Urh fort, under cover of a fierce shrapnel fire from Chair hill, across the apex of the Shuishi valley. The Japanese casualties were 120. Four officers were killed and eight wounded. The Russians left twenty dead.

CHAPTER XXXV

THE FALL OF PORT ARTHUR

General Stoessel Surrenders the Great Fortress to the Victorious Japanese on New Year's Day, 1905, after Five Days of Bloody Hand-to-Hand Fighting—Japanese Take 25,000 Prisoners, of Whom 16,000 Were in the Hospitals—Dramatic Meeting between Stoessel and Nogi—Siege Cost Nearly 80,000 Lives—The Dawn of Peace.

AT 9 O'CLOCK, January 1, 1905, Port Arthur fell; and the long siege, which cost nearly 80,000 lives and has no parallel in history outside of the siege of Troy, came to an end.

At the hour and date named above General Nogi, commander of the Japanese besieging army, received the following message from the gallant Russian defender of the Czar's "Gibraltar of the East":

"Judging by the general condition of the whole line of hostile positions held by you I find further resistance at Port Arthur useless, and for the purpose of preventing needless sacrifice of lives I propose to hold negotiations with reference to capitulation.

"Should you consent to the same, you will please appoint commissioners for discussing the order and conditions regarding capitulation and also appoint a place for such commissioners to meet the same appointed by me.

"I take this opportunity to convey to your excellency assurances of my respect. Stoessel."

Reply Made to Stoessel.

To the letter Gen. Nogi made the following reply:

"I have the honor to reply to your proposal to hold negotiations regarding the conditions and order of capitulation. For this purpose I have appointed as commissioner Maj. Gen. Ljichi, chief of staff of our army. He will be accompanied by some staff officers and civil officials. They will meet your commissioners January 2, noon, at Shuishiying.

"The commissioners of both parties will be empowered to sign a convention for the capitulation without waiting for ratification, and cause the same to take immediate effect. Authorization for such plenary powers shall be signed by the highest officer of both the negotiating parties, and the same shall be exchanged to the respective commissioners.

"I avail myself of this opportunity to convey to your excellency assurance of my respect. Nogi."

Mikado's Tribute to Stoessel.

Copies of these dispatches were hurriedly laid before the emperor, who seized the quick opportunity to pay a tribute to the courage of General Stoessel. His majesty directed Field Marshal Yamagata, chief of the general staff, to forward the following dispatch to General Nogi:

"When I respectfully informed his majesty of General Stoessel's proposal for capitulation, his majesty was pleased to state that General Stoessel has rendered commendable service to his country in the midst of difficulties, and it is his majesty's wish that military honors be shown him."

For days and weeks before this exchange of dispatches working the surrender of the great fortress, Port Arthur had been a veritable "living hell," and hand-to-hand fighting with dynamite bombs and bayonets had been a bloody daily feature of the siege.

Destruction of the Russian Fleet.

Stoessel's last hope of assistance from the Port Arthur Russian fleet vanished on August 10, when Admiral Witthoft made a sortie against Admiral Togo's fleet and lost his life and several of his most formidable ships. The Czarewitch, Admiral Witthoft's flagship, was literally shot to pieces and ready to sink when it reached the harbor of Kinchou, the German concession. The cruiser Askold, badly damaged, took refuge at Woo-Sung. Both of these vessels, together with the torpedo boats accompanying them, were compelled to dismantle and were thus put out of service. The cruiser Novik, which escaped from the battle of August 10, was overtaken by two Japanese cruisers at Saghalien island, northeast of Vladivostok, and was sunk. An exciting incident of this siege was the capture of the Russian torpedo boat destroyer Rishitelni in the harbor of Chefoo by Japanese torpedo boats. The incident raised a controversy among the powers over the right of the Japanese to enter a neutral harbor and seize a ship, but it was only a tempest in a teapot and soon blew over.

Kamimura's Victory.

On top of this disaster followed the practical destruction of the Russian Vladivostok squadron by a Japanese squadron under Admiral Kamimura. The Vladivostok squadron attempted to come to the rescue of the Port Arthur squadron, but was intercepted by Kamimura in the straits of Korea on Augusa 14. A long range fight followed, in which the Russian cruiser Rurik was sunk, and the Gromoboi and Bogatyr badly damaged. The damaged cruisers, however, succeeded in reaching Vladivostok, but were useless for fighting purposes.

The Baltic Fleet Disaster.

The combined victories of Togo and Kamimura ended the last hope of Stoessel for relief from the two Russian squadrons in far eastern waters, but a new hope was raised in his breast by the announcement that the Czar would send him a powerful fleet from the Baltic sea. The fleet, comprising some of the most powerful battleships and cruisers in the Russian navy, under command of Admiral Rojestvensky, sailed early in October. While passing through the North sea on October 21 the fleet opened fire on a fleet of British fishing vessels, killing the captain of one of the trawlers and sinking his trawl. By this incident the Baltic fleet was detained several days at Vigo, Spain, but was allowed to proceed after an international court of inquiry had been arranged for.

Before the Baltic fleet had proceeded far on its way, however, the situation at Port Arthur had become too desperate for General Stoessel to longer hold out.

The closing days of October brought notable successes to the besiegers, but they also met with serious reverses. On October 28 they gained the counterscarps of Rihlung and Sungshu forts and captured "P" fort between East Keekwan and Panlung mountains. They lost 2,000 men in this operation, and were forced to abandon the positions taken at so dear a cost.

Then came November with its furious battles, the Japanese gaining ground almost inch by inch. On November 5 and 6 the Japanese were repulsed in an attempt to carry Etse fort by storm. On the 13th the Russians in turn were repulsed in a sortie. On November 26 the Japanese began a general assault on Rihlung, Sungshu, and Keekwan forts, but although they reached the inside they were driven out with fearful loss.

Fleet Sunk at Anchor.

Then on December 6 came the capture of 203 Meter hill. From this point of vantage the Japanese bombarded the remaining Russian warships in the harbor with wonderful effect. The hill overlooked every foot of Port Arthur and the harbor. From its crest Japanese officers were enabled to direct the fire of the heavy guns beyond with such unerring aim that the Russian ships at anchor in the harbor were destroyed or sunk, one by one.

By the middle of December but one battleship, the Sevastopol, and four or five torpedo boats formed the floating remnants of Russia's once powerful fleet. The Sevastopol, driven from the harbor, took up its position in the roadstead outside the harbor, only to be destroyed by the successive attacks of Admiral Togo's torpedo flotillas.

The End at Last.

The destruction of the fleet was followed by the loss of the great forts north of the city, one by one. First Keekwan fort was captured. Then Rihlung fell on December 28. On Saturday, December 31, the Japanese captured the formidable stronghold on Sungshu mountain, and yesterday (Sunday) the forts on Panlung and Wantai mountains were captured.

That was the end. Even while Tokio was cheering the news of the capture of the forts came the news that Stoessel, after holding out six months, after losing more than half his men, after losing all of his ships, at last at bay, unable to continue the unequal struggle, had yielded at last, and the Czar's Gibraltar of the far east passed into the hands of his enemies.

The end of the siege came with dramatic suddenness. Even the Japanese, who knew that the fortress was doomed to fall within a short time, were surprised. They expected a last general assault, in which they would fight their way through gorges and over trenches and mines into the old and new towns and possibly up to the foot of Golden hill and Electric cliff.

Saturday, December 31, and Sunday, January 1, were days of furious battle. Sunday night the Japanese troops rested in their trenches for the last grand assault to be delivered in the morning.

Stoessel's Last Council of War.

That fateful Sunday afternoon General Stoessel realized that his ammunition was practically exhausted and that unless he surrendered his men would be shot down without being able to make resistance. Shortly after noon he summoned a council of his superior officers, at which Admiral Wirens represented the navy. The council met in a dugout, and shells shrieked over it incessantly as the wornout officers discussed the hopeless situation. More than one voice was choked with sobs as it gave assent to the inevitable. The agreement reached was to demand "terms of honor" or "to die fighting." The remnants of the garrison were ordered to concentrate where all of the stores had been collected, prepared to fight to the last if General Nogi proved ungenerous.

The final act in the spectacular drama really began Wednesday, December 28, when the Japanese divisions were hurled at the northern forts guarding Port Arthur in what was planned as the last general assault.

For five days without ceasing the assault was maintained. Day and night the Japanese artillery continued the fire, and assault followed assault in unending succession.

At 6 o'clock in the evening of December 31 the assault on the whole eastern side was renewed with amazing vehemence. A mine made an opening inside "H" fort, and the Japanese infantry, breaking from cover, rushed in.

The fighting lasted all night, the Japanese pushing against the whole ridge simultaneously. The Russians resisted at every point, but were slowly driven back, step by step.

New Year's day broke with the antagonists still locked in conflict, and dawn came before resistance ceased. By then the Japanese had firmly secured the Russian guns and positions on Panlung mountain, and the city was now open to them at this point, for the remaining defenses between the city and the Japanese advance post were slight and temporary.

Resistance Dies at Wantai.

At 9 o'clock Monday morning the first, center, and left column attacked the forts on Wantai mountain. The Japanese artillery was cleverly used to screen the infantry advance and by using its protection the besiegers reached the fort. But so stubborn was the resistance that not until 3:35 o'clock in the afternoon did they fully occupy the position on Wantai mountain, fighting of the fiercest character having been maintained for over six hours without ceasing.

Attacks on forts to the southeast were carried on simultaneously. Soon after midday on January 1 there was an explosion on South Keekwanshan. The Russians there immediately opened a heavy rifle fire, which soon ceased. The Japanese scouts advanced and found that the enemy was evacuating the position after blowing up the fort with its own magazine.

Russians Blow Up Forts.

The destruction of this fort on South Keekwanshan was the signal of the doom of Port Arthur. The slackening fire from all of the northern and eastern forts seemed to show that the Russian ammunition was failing.

Then all at once the Japanese realized that the beginning of the end was at hand.

Explosion after explosion came from the forts on the surrounding hills, as the Russians blew up their remaining forts.

Then the Russians began their work of destruction in the city and in the harbor. Explosion after explosion revealed the destruction of the Russian ships. They were destroyed rather than that they should fall into the hands of the victors.

The half sunken ships Retvizan, Poltava, and Pallada caught fire as the other ships were blown up inside of and near the entrance to the harbor.

At half an hour after midnight, January 2, the Russians evacuated the East Keekwanshan forts and the Japanese occupied "N" and "M" heights to the south of the fort, and at half past 12 o'clock on Monday the remaining forts were blown up by the Russians. All fighting ceased after 9 o'clock Sunday night. The siege was over.

The first definite information of the Russian intimation to surrender came at 4 o'clock in the afternoon of January 1, when the Russian envoys approached the Japanese lines south of the village of Shuishiying. They were met by a Japanese staff officer, to whom they delivered the letter of General Stoessel to General Nogi, asking to have a time for parley arranged.

Commissioners Are Named.

General Nogi replied to General Stoessel's note at once and suggested that the commissioners representing the two armies meet at the village of Shuishiying at 1 o'clock p. m. on January 2.

To this General Stoessel agreed and named as his representatives his chief of staff, Major General Reiss, Surgeon General Ballacchoff, Colonel Vostock, and two other staff officers, with two interpreters.

The Japanese commissioners were Major General Ijiohi, chief of staff; Major Yamaoka, Dr. Ariga, and two other staff officers, with two interpreters.

Major General Ijichi, with a large escort, left the Japanese headquarters at 11 o'clock Monday morning, and rode to the appointed place. Shortly before 1 o'clock the Russian commissioners arrived, accompanied by a cavalry escort.

The conference took place with few preliminaries and lasted until 9:35 o'clock at night, when the terms of capitulation were signed, as follows:

Text of the Convention Governing Port Arthur's Surrender.

Soldiers and Government Officials Prisoners.—Article 1. All Russian soldiers, marines, volunteers, also government officials at the garrison and harbor of Port Arthur are taken prisoners.

Japanese to Take All Russian War Equipment.—Article 2. All forts, batteries, warships, other ships and boats, arms, ammunition, horses, all materials for hostile use, government buildings, and all objects belonging to the Russian government shall be transferred to the Japanese army in their existing condition.

Men Garrisoning Leading Forts to Give Up First.—Article 3. On the preceding two conditions being assented to, as a guarantee for the fulfillment thereof, the men garrisoning the forts and the batteries on Etse mountain, Sunshu mountain, Antse mountain, and the line of eminences southeast therefrom shall be removed by noon of January 3, and the same shall be transferred to the Japanese army.

Providing Against Destruction of Spoils of War.—Article 4. Should Russian military or naval men be deemed to have destroyed objects named in article 2, or to have caused alteration in any way in their condition at the existing time, the signing of this compact and the negotiations shall be annulled, and the Japanese army will take free action.

Transfer of Lists of Forts, Mines, and Men.—Article 5. The Russian military and naval authorities shall prepare and transfer to the Japanese army a table showing the fortifications of Port Arthur and their respective positions, and maps showing the location of mines, underground and submarine, and all other dangerous objects; also a table showing the composition and system of the army and naval services at Port Arthur; a list of army and navy officers, with names, rank, and duties of said officers; a list of army steamers, warships, and other ships, with the numbers of their respective crews; a list of civilians, showing the number of men and women, their race and occupations.

Property to Be Disposed of by a Commission.—Article 6. Arms, including those carried on person; ammunition, war materials, government buildings, objects owned by the government, horses, warships and other ships, including their contents, excepting private property, shall be left in their present positions, and the commissioners of the Russian and Japanese armies shall decide upon the method of their transference.

Officers May Return to Russia on Parole.—Article 7. The Japanese army, considering the gallant resistance offered by the Russian army as being honorable, will permit the officers of the Russian army and navy, as well as officials belonging thereto, to carry swords and to take with them private property directly necessary for the maintenance of life. The previously mentioned officers, officials, and volunteers who will sign a written parole pledging that they will not take up arms and in no wise take action contrary to the interests of the Japanese army until the close of the war will receive the consent of the Japanese army to return to their country. Each army and navy officer will be allowed one servant, and such servant will be specially released on signing the parole.

Preparing for Assembling of Conquered Army.—Article 8. Non-commissioned officers and privates of both army and navy and volunteers shall wear their uniforms and, taking portable tents and necessary private property and commanded by their respective officers, shall assemble at such places as may be indicated by the Japanese army. The Japanese commissioners will indicate the necessary details therefor.

Sanitary Corps and Accountants to Be Retained.—Article 9. The sanitary corps and the accountants belonging to the Russian army and navy shall be retained by the Japanese while their services are deemed necessary for the caring for sick and wounded prisoners. During such time such corps shall be required to render service under the direction of the sanitary corps and accountants of the Japanese army.

Treatment of Residents to Be Determined Later.—Article 10. The treatment to be accorded to the residents, the transfer of books and documents relating to municipal administration and finance, and also detailed files necessary for the enforcement of this compact shall be embodied in a supplementary compact. The supplementary compact shall have the same force as this compact.

Compact Goes into Immediate Effect.—Article 11. One copy each of this compact shall be prepared for the Japanese and Russian armies, and it shall have immediate effect upon signature thereof.

Nearly Twenty-five Thousand Prisoners.

Pursuant to the terms of capitulation the Russian troops marched out of Port Arthur at noon on January 5, stacked their arms and became prisoners of war. The Japanese troops marched into the city, but without making any display to humiliate their captured foes.

The transfer of prisoners at Port Arthur was completed at 4:30 p. m. Saturday. The total number of officers transferred was 878, men 23,491.

General Nogi's report shows the transfers were as follows:

	Officers.	Men.
General Stoessel's Kwantung command	9	39
General Stoessel's headquarters	6	15
Engineers' company	11	269
Telegraph corps	4	61
Railroad detachment	1	55
Cavalry	4	177
Officers and crews of warships—		
Retvizan	22	...
Pobieda	22	...
Pallada	11	...
Peresviet	15	...
Poltava	16	311
Sevastopol	31	507
Bayan	15	259
Bobr	12	99
Storozhovoi	4	52
Otvashni	6	124
Gilyak	5	72
Amur	7	173
Naval defense headquarters	3	320
Harbor office	60	925
Naval brigade	59	31
Torpedo brigade	10	142
Field posts and telegraphs	33	23
Totals	366	3,654

Three judges and three clerks of courts were also transferred.

Of the 24,369 officers and men captured about 16,000 were sick and wounded. The Russian and Japanese medical staffs worked together in the hospitals caring for the invalided soldiers and sailors.

Stoessel and Nogi Meet.

The meeting of General Nogi and General Stoessel was as dramatic as the conclusion of the siege. It had previously been arranged to take place at noon in the single undamaged house of the village of Shuishi. This house was a miserable hovel called plum tree cottage.

Through a misunderstanding General Stoessel rode out to Port Arthur at 10 o'clock, accompanied by Colonel Reiss and two staff officers, to the Japanese lines, but missed the Japanese officer delegated to escort him to the meeting place. The general rode there without an escort and was received by a junior officer who happened to be on the spot. The latter telephoned to Nogi, who hurried his departure from headquarters and arrived at 11 o'clock, accompanied by Major General Ijichi, his chief of staff, and Colonels Yasuhara, Matsudaira and Watanabe, staff officers, and M. Kawskarin, secretary of the foreign office at Tokio.

Nogi Offers Mikado's Message.

When Nogi, looking careworn, entered the compound of the cottage, the two generals cordially shook hands and Nogi through an interpreter expressed his pleasure at meeting a general who had fought so bravely and gallantly for his emperor and country.

General Stoessel thanked General Nogi for the pleasure of meeting the hero of the victorious army.

General Nogi explained that he had received a message from his emperor asking that the greatest consideration be shown to General Stoessel and his officers in appreciation of their splendid loyalty to their emperor and country. Because of that wish, he added, the Russian officers would be allowed to wear their swords.

Stoessel Expresses His Gratitude.

General Stoessel expressed his gratitude to the Japanese emperor for thus saving the honor of Stoessel's family and said his descendants would appreciate the thoughtful kindness of the emperor of Japan. The general also expressed the gratitude of his officers and thanked Nogi for sending the message from Stoessel to Emperor Nicholas and transmitting his majesty's reply, which reads as follows:

"I allow each officer to profit by the reserved privilege to return to Russia under the obligation not to take further part in the present war or to share the destinies of their men."

Praises the Japanese Gunners.

Both generals then mutually praised each other and their officers for their bravery. The conversation afterward turned on the explosion of the mine at Sungshu mountain fort. General Stoessel said the entire garrison of the fort was killed or made prisoners.

The Russian commander greatly praised the Japanese artillery shooting, especially the concentrated fire instantaneously with the explosion of Sungshu mine. The gallant deeds of the Japanese infantry, General Stoessel added, spoke for themselves. It was impossible to exaggerate their good qualities. The skillful work of the engineers had also won his admiration.

Regrets Death of Nogi's Sons.

Continuing, General Stoessel said he had heard that General Nogi had lost both his sons and praised his loyalty in this sacrificing his sons, who had died fighting for their emperor and country.

General Nogi smilingly replied: "One of my sons gave his life at Nanshan and the other at 203 Meter hill. Both of these positions were of the greatest importance to the Japanese army. I am glad that the sacrifice of my sons' lives was in the capture of such important positions, as I feel the sacrifices were not made in vain. Their lives were nothing compared to the objects sought."

The Japanese commander requested General Stoessel to continue to occupy his residence at Port Arthur until arrangements were completed for the return of himself and family to Russia.

Referring to the burial of the dead, General Nogi said the Japanese, since the beginning of the military operations, had always buried the Russian dead. Those found later would be interred at a special spot and a suitable memorial would be erected as a tribute to the bravery of Japan's former foes.

Both are Photographed Together.

After luncheon at which both generals sat together a group photograph was taken at the cottage and General Stoessel remounted his charger to show the horse's good points, said good-by to General Nogi and rode back to Port Arthur. The quiet and even solemn meeting of the generals ended at about 1 o'clock.

The regular Russian soldiers in Port Arthur marched out Jan. 5. The only troops left in the city were the volunteers.

Two fires were started in Port Arthur during the day for which General Stoessel apologized. He said the volunteers were unable to control the populace and he desired that the Japanese enter Port Arthur immediately to keep order.

The formal entry of the Japanese army into Port Arthur took place on Sunday, Jan. 8.

Praise Gallant Defense.

The gallant defense of Stoessel and his men has nowhere been given a finer appreciation than in the land of his foes, and Japan gladly embraced the opportunity to show her magnanimity and admiration of the gallantry of Port Arthur's defenders by allowing them all the honors which war permits a victorious army to bestow upon the vanquished.

The siege and defense of Russia's stronghold in the far east were marked by bravery, gallantry and desperation unequaled in modern warfare and hardly excelled in military history. The story of the operations around Port Arthur is one of repeated fighting, both by land and sea, of the most desperate and thrilling character. Isolated instances of heroism that would have set the world ringing under less overwhelming circumstances have been dwarfed by the generally magnificent conduct of both forces.

Daring by Land and Sea.

By sea there were torpedo-boat dashes of superb recklessness and big ships have ploughed through mine fields with heroic disregard to give battle or in wild efforts to escape. By land the Japanese hurled themselves against positions declared to be impregnable. They faced and scaled rocky heights crowned with batteries and crowded with defenders, suffering losses that military experts say would have appalled any European army.

In the doomed fortress its people lived under a devastating rain of shell and shrapnel. On scanty rations, besieged on every side, knowing that hope of succor or escape was vain, the garrison fought with a stubbornness that evoked the admiration of the world. They met the untiring assaults of the Japanese with a grim valor that won even the praise of their foe, and the fighting has been waged with a relentlessness than often refused truces to bury the dead and collect the wounded.

The Dawn of Peace.

The capture of Port Arthur marks the final chapter in the history of the war. The Russian land forces were left cooped up in Mukden and held in check by a powerful army under Field Marshal Oyama. By the fall of Port Arthur Japan was enabled to concentrate her entire military force against General Kuropatkin, and thus render futile Russia's efforts to continue the war with any hope of success.

The fall of Port Arthur was the decisive event of the year's struggle, and as the flag of the Rising Sun was raised over its shell-torn ramparts the nations of the world recognized it as an emblem that heralded the dawn of peace.

Related Titles from Westphalia Press

The Limits of Moderation: Jimmy Carter and the Ironies of American Liberalism
by Leo P. Ribuffo

The Limits of Moderation: Jimmy Carter and the Ironies of American Liberalism is not a finished product. Yet, this book is a close and careful history of a short yet transformative period in American political history, when big changes were afoot. and continue to shape our world.

Bunker Diplomacy: An Arab-American in the U.S. Foreign Service
by Nabeel Khoury

After twenty-five years in the Foreign Service, Dr. Nabeel A. Khoury retired from the U.S. Department of State in 2013 with the rank of Minister Counselor. In his last overseas posting, Khoury served as deputy chief of mission at the U.S. embassy in Yemen (2004-2007).

Energy Law and Policy in a Climate-Constrained World
by Victor Byers Flatt, Alfonso López de la Osa Escribano, Aubin Nzaou-Kongo

This book presents reflections on concepts, foreign policy, regional and international cooperation, and the specific role the state is to play when it comes to such thing as energy law and policy.

The Zelensky Method
by Grant Farred

Locating Russian's war within a global context, The Zelensky Method is unsparing in its critique of those nations, who have refused to condemn Russia's invasion and are doing everything they can to prevent economic sanctions from being imposed on the Kremlin.

The Lord of the Desert: A Study of the Papers of the British Officer John B. Glubb in Jordan and Iraq
by Dr. Sa'ad Abudayeh

John Bajot Glubb, a British engineer officer, was sent to Iraq in 1920 to resolve the problems which erupted after the Iraqi revolt. He remained in the area for ten years, working with the Bedouins. In 1930, he moved to Jordan for twenty-six successful years. He invented what Dr. Abudayeh calls the Diplomacy of Desert.

Managing Challenges for the Flint Water Crisis
Edited by Toyna E. Thornton, Andrew D. Williams, Katherine M. Simon, Jennifer F. Sklarew

This edited volume examines several public management and intergovernmental failures, with particular attention on social, political, and financial impacts. Understanding disaster meaning, even causality, is essential to the problem-solving process.

Resistance: Reflections on Survival, Hope and Love
Poetry by William Morris, Photography by Jackie Malden

Resistance is a book of poems with photographs or a book of photographs with poems depending on your perspective. The book is comprised of three sections titled respectively: On Survival, On Hope, and On Love.

Abortion and Informed Common Sense
by Max J. Skidmore

The controversy over a woman's "right to choose," as opposed to the numerous "rights" that abortion opponents decide should be assumed to exist for "unborn children," has always struck me as incomplete. Two missing elements of the argument seems obvious, yet they remain almost completely overlooked.

Geopolitics of Outer Space: Global Security and Development
by Ilayda Aydin

A desire for increased security and rapid development is driving nation-states to engage in an intensifying competition for the unique assets of space. This book analyses the Chinese-American space discourse from the lenses of international relations theory, history and political psychology to explore these questions.

The Athenian Year Primer: Attic Time-Reckoning and the Julian Calendar by Christopher Planeaux

The ability to translate ancient Athenian calendar references into precise Julian-Gregorian dates will not only assist Ancient Historians and Classicists to date numerous historical events with much greater accuracy but also aid epigraphists in the restorations of numerous Attic inscriptions.

The Politics of Fiscal Responsibility: A Comparative Perspective by Tonya E. Thornton and F. Stevens Redburn

Fiscal policy challenges following the Great Recession forced members of the Organisation for Economic Co-operation and Development (OECD) to implement a set of economic policies to manage public debt.

China & Europe: The Turning Point by David Baverez

In creating five fictitious conversations between Xi Jinping and five European experts, David Baverez, who lives and works in Hong Kong, offers up a totally new vision of the relationship between China and Europe.

Issues in Maritime Cyber Security Edited by Dr. Joe DiRenzo III, Dr. Nicole K. Drumhiller, and Dr. Fred S. Roberts

The complexity of making MTS safe from cyber attack is daunting and the need for all stakeholders in both government (at all levels) and private industry to be involved in cyber security is more significant than ever as the use of the MTS continues to grow.

Freemasonry, Heir to the Enlightenment by Cécile Révauger

Modern Freemasonry may have mythical roots in Solomon's time but is really the heir to the Enlightenment. Ever since the early eighteenth century freemasons have endeavored to convey the values of the Enlightenment in the cultural, political and religious fields, in Europe, the American colonies and the emerging United States.

Made in the USA
Monee, IL
04 November 2026

67386643R00252